General Chemistry
Laboratory Manual

Edited by
Edward Witten
Northeastern University

KENDALL/HUNT PUBLISHING COMPANY
4050 Westmark Drive   Dubuque, Iowa 52002

Copyright © 2006 by Northeastern University Chemistry Department

ISBN 978-0-7575-3220-7

Printed in the United States of America
20 19 18 17 16 15 14

Contents

BOILING POINT DETERMINATION

Introduction

This experiment investigates a characteristic physical property, namely the *boiling point*. In future experiments you will study additional physical properties such as *melting point* and *density*. You also will become familiar with the separation technique known as filtering. Physical properties can be used to identify a substance. For example, if you have prepared a white solid, you can determine its physical properties (density, melting point, boiling point, solubility, etc.) and compare these properties to known substances in the Handbook of Chemistry and Physics. If the boiling points, melting points, densities, etc., of the two samples are identical, then the two samples are almost certainly the same substance. The more physical properties measured the more likely the match.

Today's experiment is especially interesting since you will be measuring the boiling point of an unknown liquid using only a very *small amount* of the liquid: This sort of "micro" technique can be very useful in situations where a researcher may only have available a *few drops* of some new substance. It serves as a good introduction to your lab work in the General Chemistry course you are taking.

A second advantage of using this "micro" technique is that the amount of chemical needed and the corresponding waste generated is greatly reduced.

Before coming to the laboratory to perform this experiment, you should read over the sections in your textbook that cover the topics of boiling point and vapor pressure.

The **BOILING POINT** of a substance at a given pressure is a **CHARACTERISTIC** property of the substance, which may be used as a point of **IDENTIFICATION**. The boiling point of a substance is defined as *the temperature at which the pressure of the vapor of the substance becomes equal to the prevailing external pressure above the substance.* In the case of a system **OPEN TO THE ATMOSPHERE** in the lab, the prevailing external pressure is just the lab's **BAROMETRIC PRESSURE (atmospheric pressure).** The boiling point of a liquid is thus a function of the prevailing pressure above the liquid. Changing the pressure above the liquid changes the boiling point of the liquid. For example, at sea level water boils at 100°C if the atmospheric pressure is 760 mm Hg (30 in Hg). However, in Denver, where the typical atmospheric pressure is closer to about 660 mm Hg (26 in Hg), the boiling point of water is about 96°C.

Since a liquid can boil at many temperatures, scientists have defined a reference pressure of 1 atmosphere (760 mm Hg or 30 in Hg) to measure the boiling point of a liquid. The temperature called the **normal boiling point** is defined *as the temperature at which a liquid boils when the external pressure (pressure above a liquid) is exactly 760 mm Hg.* This allows one to compare boiling points of various substances at a given reference point.

In order for a liquid to boil easily, there must be some place where *gas bubbles can form (nucleate)* easily; without this, "*bumping*" or "*superheating*" of the liquid can occur, leading to incorrect results or loss of the liquid being heated. When bulk amounts of liquids are to be boiled in the lab (such as in a water bath), **BOILING STONES** are used for nucleation. (Note: One should put only two or three boiling chips into the liquid being heated, not a handful of twenty to thirty.)

In this experiment, where we are dealing with a very small amount of liquid, a small broken **CAPILLARY TUBE** provides a rough surface upon which gas bubbles can form more easily than on the smooth walls of the container.

The instrument used in the laboratory for temperature measurement is, of course, the **THERMOMETER**. Before a thermometer can be considered *reliable* in its readings, it should be **CALIBRATED AT SOME REFERENCE TEMPERATURES**. That is, several systems of absolutely *known* temperature should be measured with the thermometer. If the thermometer is in error, one can draw a **CALIBRATION CURVE,** which will allow you to make corrections to the observed temperature reading on your thermometer. Applying this correction to the observed temperature reading on your thermometer will allow you to determine the actual temperature.

This curve can be used subsequently to determine the error likely in any given reading. For example, if your thermometer reads one degree too high at the boiling point of water, one would subtract one degree from the reading to get the correct temperature. You will calibrate your thermometer at the freezing point of water (by immersing it in an ice water slushy mixture whose temperature is exactly 0°C) and at the boiling point of water (whose temperature is near 100°C, but whose exact temperature depends on the days atmospheric pressure).

We can define the error of the thermometer to be *the observed temperature reading minus the correct temperature*. The correction can be defined as *the number of degrees that must be added or subtracted from the observed value to get the correct value*. For example, on a given day the atmospheric pressure is 770 mm Hg and the boiling point of water at this pressure is 100.4°C. When we measure the boiling point using the thermometer provided, the temperature read 102.8°C.

The error would be 102.8° - 100.4°C = 2.4°C

The correction would be to subtract 2.4°C from the observed reading or -2.4°C.

Since the thermometer is reading 2.4°C too high, it has an error of +2.4°C and the required correction is -2.4°C (meaning that we need to subtract 2.4°C from the observed value.)

ANSWER THE PRE-LAB QUESTIONS BEFORE COMING TO THE LAB EACH WEEK.

CHEMICAL INFORMATION

The unknowns used in this experiment are one of the following substances: methanol, ethanol or isopropyl alcohol (rubbing alcohol).

Methanol—flammable liquid, toxic by ingestion
Ethanol—flammable liquid, toxic by ingestion
Isopropyl alcohol—(rubbing alcohol) flammable liquid, toxic by ingestion

PROCEDURE

WEAR SAFETY GLASSES AT ALL TIMES!!

(1) Calibration of the Thermometer

Fill a 150 mL or 250 mL beaker with ice and add water to make a slushy mixture (about 80% ice and 20% water). Too much water will cause your solution to take too long to reach thermal equilibrium. **GENTLY** stir this slushy bath with your thermometer until the temperature reading **NO LONGER CHANGES**. Do **NOT** let the thermometer **TOUCH THE GLASS** of the container (or you will not be reading the temperature of the ice bath). Your thermometer must be read **WHILE IN THE ICE BATH** (it is **NOT** like a body thermometer, whose reading doesn't change unless it is shaken).

Record the thermometer reading to the nearest $0.2^{\circ}C$ (i.e., two-tenths of a degree, estimating BETWEEN the marks on the thermometer barrel). Allow the thermometer to warm up in a safe place on the bench.

If a rubber stopper is not already attached to your thermometer, use the following procedure to safely place a rubber stopper onto your thermometer. Using **GLYCERINE** as a **LUBRICANT**, and **PROTECTING YOUR HANDS WITH A TOWEL**, insert the **TOP** of your thermometer about 1 inch through a rubber stopper in such a way that the thermometer can still be read from 50 to $100^{\circ}C$. (You only need to do this if your thermometer is not already attached to a rubber stopper.)

Obtain a 400 mL or a 600 mL beaker from your lab locker and fill it about 3/4 full with water from the sink.

Obtain a ring stand and a large ring from the area above your work space. Attach the ring to the ring stand. Find the wire gauze in your lab locker and place it on top of the ring. This will give more support to your beaker and also prevent the bottom of the beaker from becoming charred. Place the beaker filled with water on the wire gauze. Attach an adjustable clamp to the ring stand.

Suspend the thermometer in the water, using a clamp to hold the rubber stopper, again taking care that the thermometer does not touch the walls or bottom of the container.

Light your Bunsen burner with your burner lighter (sparker) and allow the water to boil. Wait until your temperature reading no longer changes, and then record the reading to the nearest 0.2°C in your lab notebook.

Your lab instructor will write the day's atmospheric pressure and its corresponding boiling point of water on the blackboard. Record this information in your lab notebook.

If necessary draw a calibration curve (this should be done if your thermometer shows any errors. Your lab instructors will go over during the pre-lab talk how to draw your calibration curve. The calculation section of this experiment will also describe how to draw a calibration curve.

(2) Boiling Point Determination

Obtain a small test tube from the table or hood where lab supplies are located in your lab room.

Obtain a 600 mL beaker from your locker.

Inspect the sample apparatus for assembling your apparatus **CORRECTLY**.

Connect your small test tube to the thermometer by using two rubber bands. (A rubber stopper should be on your thermometer; if not, attach one.)

Obtain a liquid unknown and record the unknown number in your lab notebook. Add sufficient unknown liquid to the test tube to give a liquid column height of about 4 cm (save the rest of the unknown for a repeat determination).

Break a 2 cm length from the **SEALED** end of a small **CAPILLARY TUBE**. Insert the small piece of capillary **SEALED END UP** in the tube of unknown liquid.

Fill your 600 mL beaker with **COLD** water, making sure that there is enough water in the beaker to **COVER** all of the unknown liquid in the sample tube. Be sure NOT to allow water to enter the **TOP** of the test tube (if water mixes with the unknown sample, erroneous results will be obtained).

Immerse the unknown liquid/thermometer apparatus, and then heat the water bath with frequent stirring until bubbles rise in a **STEADY STREAM FROM THE SMALL CAPILLARY**. (The water at this point should **NOT BE BOILING**: If the water is boiling, you have **HEATED TOO MUCH!**) It is likely that you will have boiled away all of your unknown. If this happens, you need to start this process again.

When bubbles have risen steadily from the small capillary for several seconds, REMOVE THE HEAT, stir the water bath **CONSTANTLY,** and record the temperature at the **INSTANT WHEN THE BUBBLES STOP COMING FROM THE CAPILLARY**. (At this point, the pressure of **VAPOR** inside the small capillary is equal to the atmospheric pressure.) **This is your boiling point.**

NOTE: When the last bubble comes out of the capillary tube, the unknown liquid will start to fill up the capillary tube. Watch to see if you can observe this happening.

Record the temperature when the bubbling stops to the nearest 0.2°C: This is the boiling point.

To check on your measurements, you will **REPEAT** the determination of boiling point. Refill the beaker with **COLD** water. If necessary, add more unknown liquid to the sample tube to preserve the 4 cm liquid height. The broken capillary used for the first determination will have filled with liquid as the temperature dropped, and cannot be reused (the original capillary does **NOT** have to be removed, however). Add a **SECOND** broken capillary tube to the unknown liquid sample, sealed end up.

Repeat the determination of the boiling point, following the same procedure as above. **YOUR SECOND DETERMINATION OF THE BOILING POINT SHOULD CHECK WITH THE FIRST DETERMINATION WITHIN TWO DEGREES: IF NOT, REPEAT UNTIL YOU GET TWO VALUES THAT AGREE.**

Return your unknown vial to the location from which it was obtained (probably a yellow or blue bin or a plastic tray).

Discard your unknown liquid from your small test tube into the bottle labeled **WASTE EXPERIMENT 1 BOILING POINT UNKNOWNS**.

Throw all glass to be discarded into the broken glass box located in your lab.

Wash your hands thoroughly before leaving the lab.

REMEMBER TO TURN IN THE YELLOW COPY OF YOUR LAB NOTEBOOK TO YOUR TA.

DRAWING A CALIBRATION CURVE

1. Assign the actual measured temperature to the x-axis.
2. Assign the error (or correction) to the y-axis.
    The error is how much your experimental determined temperature is off from what the actual temperature is. For example, your thermometer reads 2.6°C (at the freezing point of water, where the actual freezing point of water is 0.0°C). The error is +2.6°C. Your temperature is 2.6°C too high! The correction is what action you must take to get the correct temperature. In this example, one must subtract 2.6°C from the experimentally determined temperature. The correction would thus be -2.6°C.
3. Plot the experimentally determined temperature on the x-axis and the error (or correction) on the y-axis. Your graph should now have two dots, one for the boiling point and one for the freezing point.
4. Connect the two points with a straight line.
5. To read your graph, locate your experimentally determined boiling temperature on the x-axis. Draw a straight line up to your calibration curve. Lastly draw a straight line from your calibration curve over to the y-axis. You can now determine the error at that experimentally measured temperature.
6. Apply the error to your experimentally determined boiling point and report it on your report sheet.

EXAMPLE   Plot of Actual Temperature vs. Correction

Barometric pressure is 767 mm Hg. The boiling point of water at this pressure is 100.2°C.

Thermometer reading at 0°C is 1.3°C.

Thermometer reading at the boiling point of water is 103.6°C.

AT THE FREEZING POINT OF WATER

Thermometer error is +1.3°C and the correction is -1.3°C.

AT THE BOILING POINT OF WATER

Thermometer error is +3.4°C and the correction is -3.4°C.

Plot one point on your graph paper at
        X axis = 1.3°C and Y axis = -1.3°C
Plot the second point on your graph paper at
        X axis is 103.6°C and Y axis = -3.4°C
Connect these two points and your graph is now done.

Boiling Point Determination
Report Sheet (37 Points)

Name _____ Date _____

Lab Instructor _____ Lab Day _____

I.   <u>Thermometer Calibration</u>

Barometric Pressure   _____

Correct boiling point of water   _____

Thermometer reading in ice water bath   _____

Thermometer reading in boiling water bath _____

Thermometer errors: at 0°C   _____

                           at 100°C   _____

Thermometer corrections: at 0°   _____

                           at 100°C   _____

II   <u>Unknown Determination</u>

Code number of unknown

Observed boiling point   _____   Corrected boiling point   _____

                           _____   _____

                           _____   _____

Boiling Point Determination
Post-Lab Questions (33 Points)

Name _____ Date _____

Lab Instructor _____ Lab Day _____

Questions 1 through 8 are worth 3 points each. Questions 9 through 17 are worth 1 point each.

1.   Why is it necessary to calibrate your thermometer?

2.   Why is it necessary to know the atmospheric pressure in order to calibrate your thermometer?

3.   Why do you need to make an ice water slush mixture (instead of only ice) in order to calibrate your thermometer at 0°C?

4.   What function does the inverted capillary tube perform?

5.   A student calibrated his/her thermometer and obtained a reading of -2.0°C in an ice water slush bath. What is the error (including the sign) in the thermometer reading?

6. Rubbing alcohol boils at 82.4°C at a barometric pressure of 760 mm Hg. If the barometric pressure in a laboratory was 720 mm Hg, would the observed boiling point of rubbing alcohol be higher or lower than 82.4°C?

7. Describe the proper technique for inserting a thermometer into a rubber stopper.

8. What are the four pieces of safety equipment present in each lab? Give a brief description of how and when to use them.

Answer Questions 9 through 17 either TRUE or FALSE

9. Safety glasses should be worn at all times while in the laboratory regardless of whether you are working on your experiment or not.

     _____

10. To determine the safety of a chemical read the label.

     _____

11. Bunsen burners should never be left unattended.

     _____

12. Never inhale chemical vapors.

     _____

13. One is allowed to work in the chemistry laboratory before your lab instructor is present.

     _____

14. Dispose of all chemicals as directed by your lab instructor.

     _____

15. Do not fool around in the laboratory.

     _____

16. Broken glass should be put in the broken glass boxes in your lab.

     _____

17. Clean up solid and liquid spills immediately.

     _____

Boiling Point Determination
Preliminary Questions (10 Points)

Name _____ Date _____

Lab Instructor _____ Lab Day _____

1. Define boiling point.

2. How will an increase in pressure affect the boiling point of a substance?

3. What two reference points will you use to calibrate your thermometer?

4. After finishing your experiment, what is the proper procedure for disposing the remaining unknown in your small test tube?

5. Look up in your textbook or another reference source the boiling points of the three different unknowns used in this experiment.

    Methanol _____   Ethanol _____ Isopropyl alcohol _____

6. Draw the calibration curve described on page 6.

11

Recrystallization and Melting Point Determination

Introduction

RECRYSTALLIZATION is a standard and efficient process for PURIFICATION of solid substances. In this process, the IMPURE solid is first dissolved in HOT water. INSOLUBLE CONTAMINANTS can then be removed by GRAVITY FILTRATION of the hot solution through ordinary filter paper. The solution that passes through the filter, called the FILTRATE, is then cooled to room temperature (or below) to permit CRYSTALLIZATION of the desired material. Water is then removed from the crystals of the desired material by SUCTION FILTRATION. The damp crystals are then dried in the air over several days to remove residual water.

The MELTING POINT is a CHARACTERISTIC property of a pure crystalline substance and may be used as a point of IDENTIFICATION for the substance. PURE substances usually melt SHARPLY (or over a very narrow range of temperatures).

The presence of a dissolved IMPURITY almost always has TWO EFFECTS on the melting point: The MIXTURE of chemicals (major substance plus impurity) tends to melt at a LOWER TEMPERATURE INITIALLY, and over a much BROADER RANGE OF TEMPERATURES overall.

In this experiment we determine accurately the melting point of a recrystallized UNKNOWN chemical, and then identify the unknown by COMPARING its melting point with the melting points of KNOWN chemicals. A MIXTURE of the UNKNOWN chemical is then made with the KNOWN chemical it is believed to be, and the melting point of the mixture is determined. If the known and unknown chemicals are IDENTICAL, the mixture should melt at the SAME TEMPERATURE. If the known and unknown are different, the melting point of the mixture should show a large depression (10°C or greater) and will melt over a broad temperature range.

Summary

An unknown substance is purified by recrystallization from water and is allowed to dry. Its melting point is then determined, and its identity confirmed by mixed melting point.

Supplies

Capillary tubes; rubber bands

CHEMICALS

Acetanilide, toluic acid, benzoic acid, p-toluene sulfonamide and salicylic acid are all organic flammable solids.

Acetanilide—irritant
Benzoic acid—irritant, toxic by ingestion
p-Toluene sulfonamide—irritant
o-Toluic acid—irritant, toxic by ingestion
Salicylic acid—toxic by ingestion

CAUTION! WEAR SAFETY GLASSES AT ALL TIMES!!

Procedure

(1) Recrystallization

Obtain a melting point unknown from the lab and **RECORD THE UNKNOWN NUMBER** in both your lab notebook and on the Experiment 2 report sheet.

Inspect the sample apparatus for hot GRAVITY filtration and for SUCTION filtration.

Weigh an empty piece of filter paper or weighing paper. Transfer your unknown to the filter or weighing paper and reweigh.

Transfer your unknown to a 250 mL Erlenmeyer flask and add about 100 mL of water.

Return your empty plastic container to the tray you took it from before you leave the lab.

Heat the water to boiling and stir the mixture occasionally with a glass rod during the solution process. (The insoluble impurity, carbon, will not dissolve.)

Filter the hot solution of the unknown through a filter paper in the gravity funnel into a 250 mL beaker. CAUTION: YOU WILL NOT BE ABLE TO HANDLE THE FLASK WITH YOUR BARE HANDS. YOU WILL NEED TO USE YOUR CRUCIBLE TONGS OR SEVERAL PIECES OF PAPER TOWEL.

Dispose of your filter paper into the beaker labeled used filter paper.

Heat the solution in your 250 mL beaker until the volume is reduced to between 40 and 50 mL. Cool the filtrate in the beaker by immersion in an ice/water bath.

When crystallization is complete (the filtrate reaches 0°C), filter the crystals by suction, using a water aspirator, suction flask, Buchner funnel, and adapter. Remove as much of the liquid as possible by suction (press down on the Buchner funnel with the palm of your hand to ensure a tight seal). Make sure you take the proper size filter paper to fit your Buchner funnel. It should cover all the holes and be flat on the bottom. If it goes up the side of the funnel, the filter paper is too big for your funnel.

Weigh a clean empty watch glass and record its mass in your lab notebook.

Transfer the moist crystals to a clean watch glass and allow them to dry until the next lab period. Before you leave the lab, place the watch glass flat on top of your 600 mL beaker in your locker.

Dispose of your filter paper in the Buchner funnel in the beaker labeled used filter paper.

Dispose of the liquid in your filter flask down the drain.

(2) Melting Point Determination

Weigh your watch glass with the crystals.

There are two methods for determining the melting point of your unknowns. The first is using an electronic device called a Digital Melting Point Apparatus. The second is to use a mineral oil bath to melt your crystals. Directions for using both are listed below.

DIRECTIONS FOR USING DIGITAL MELTING POINT APPARATUS

1. Your lab instructor will show you how to fill the melting point capillary. The column of crystals in the bottom of the capillary tube should only be about 1 cm high. Place the capillary tube in the back of the eyepiece.
2. Turn on the apparatus. It is the black switch on the back left of the transformer. Do not change the settings on the red switch.
3. The device will beep and should read between 20°C and 25°C.
4. Hit the up arrow ▲ symbol to set the desired temperature. (The machine moves temperature upward in 10°C increments.) Push button for every 10°C increase desired. The symbol is located on the bottom of the control area (next to the down symbol and the clear symbol).
   ▲
5. When the desired set temperature is reached push the ⌐ symbol. This starts the instrument heating.
6. When it reaches the set temperature, the instrument will beep three times. If your crystals have not melted, increase the set temperature.
8. To increase the set temperature, push the clear symbol and the push the up arrow symbol until the new set temperature is reached and shows in the window.
   ▲
9. Push the ⌐ symbol to start the unit heating again to attain the new set temperature.
10. The recommended procedure is to set the device initially to 100°C. If your crystals have not melted by this temperature, increase the temperature by 10°C and reheat. Use 10°C increases until the crystals melt.

11. After finishing your melting point, it will be necessary to cool down the instrument to below 100°C before the next person can melt their crystals.

12. To cool down the instrument, push the clear symbol and then the down ▼ symbol until the desired temperature is reached (probably you should set it to 90°C for the next person). Note the apparatus changes temperature settings downward in 1°C increments.

MIXED MELTING POINT DETERMINATION

From the stockroom obtain a small sample of the pure compound you believe your unknown to be. Mix together THOROUGHLY roughly equal amounts of your recrystallized unknown and the known compound in the plastic container that you obtained from the stockroom that contains the known compound.

Determine the melting range of the mixture carefully. If there is no significant melting point depression, the identity of your unknown is confirmed. If you observe a large depression in the mixed melting point, try a different substance and repeat this procedure. Continue trying different substances until you find a substance that does not depress the melting point.

Dispose of your crystals in the WASTE EXPERIMENT 2 CRYSTALS beaker.
Dispose of your plastic boats into the plastic boats beaker.
Dispose of your capillary tubes into the capillary tubes beaker.

REMEMBER TO TURN IN THE YELLOW COPY OF YOUR LAB NOTEBOOK TO YOUR TA.

POSSIBLE UNKNOWNS AND MELTING RANGES

o-Toluic acid	100–106°C
Acetanilide	110–117°C
Benzoic acid	120–125°C
P-Toluene sulfonamide	133–140°C
Salicylic acid	150–155°C

Recrystallization and Melting Point Determination Report Sheet (30 Points)

Name _____   Date_____

Lab Instructor _____   Lab Day _____

Report

(1) Unknown number _____

Mass of filter or weighing paper   _____ g

Mass of unknown and paper   _____ g

Mass of unknown   _____ g

Mass of empty watch glass   _____ g

Mass of watch glass and crystals   _____ g

Mass of crystals recovered   _____ g

Percent recovery   _____ g

Approximate melting point   _____ °C

Accurate melting point (range)   _____ °C

(2)   Mixed melting point of unknown with _____ melted at _____ °C

Mixed melting point of unknown with _____ melted at _____ °C

(3) The recrystallized unknown substance is _____

Recrystallization and Melting Point Determination
Post-Lab Questions (40 points)

Name _____ Date_____

Lab Instructor _____ Lab Day _____

1. Describe the process of recrystallization (not a definition).

2. Why does one boil the charcoal–unknown–water mixture as opposed to using only cold water?

3. Why is it necessary to wait until the following week before determining the melting point of your unknown?

4. What is a mixed melting point? How is it used to confirm the identity of an unknown substance?

5. Describe the process of suction filtration and explain why you might use this technique.

6.   Calculate the percent recovery of your purified crystals.

Recrystallization and Melting Point Determination Preliminary Questions (10 Points)

Name _____ Date _____

Lab Instructor _____ Lab Day _____

(Show calculations on the back of the page.)

1. Define solubility

2. A substance has a solubility of 12.5 g per 100 mL of solvent at 100°C and 2.45 g per 100 mL solvent at 10°C. How many grams of the substance can dissolve in 25 mL of solvent at 100°C? How many grams of the substance can dissolve in 25 mL of solvent at 10°C?

    _____ grams can dissolve in 25 mL at 100°C

    _____ grams can dissolve in 25 mL at 10°C

3. What hazards are associated with the unknowns?

4. What is the proper disposal of all filter paper used in this experiment?

5. A student began with a 3.12 g sample that contained their unknown and charcoal. After recrystallization the mass of dried crystals recovered was 2.61 g. What was the percent recovery?

Density Determinations

Introduction

In general, DENSITY is defined as the MASS of substance per unit VOLUME, or

$$\text{Density} = \frac{\text{Mass}}{\text{Volume}} \qquad \text{or} \qquad D = \frac{M}{V}$$

This definition and formula are true whether we are considering PURE SUBSTANCES (for example, a pure metal, such as iron), or SOLUTIONS of some solute in a solvent such as water (as we do in this experiment with a solution of sodium chloride). The RATIO of mass to volume, regardless of the nature of the substances involved, gives the density. In this experiment, we will determine each of these quantities (mass and volume) INDEPENDENTLY of each other, and then we will calculate the RATIO—the DENSITY.

For a PURE SUBSTANCE, the density at a given temperature is CHARACTERISTIC of the substance. For a solution, knowing the density and the concentration of the solution permits calculation of the amount of solute in a given amount of solution.

The first part of this experiment involves measuring the density of some irregularly shaped chunks of rock. Determining the MASS of the chunks is done directly with the lab balances. However, determining the VOLUME of the chunks of rock is somewhat less direct. We make use of Archimedes Principle: that is, an insoluble solid will DISPLACE a volume of water equal to its own volume. By adding the chunks of rock to a measured quantity of water in a graduated cylinder, we can determine the volume of the rock chunks by the EXTENT TO WHICH THE WATER LEVEL IN THE CYLINDER IS CHANGED BY THE ADDITION OF THE ROCKS.

The second part of the experiment involves first PREPARING a particular solution, and then MEASURING ITS DENSITY by determining the mass of a particular volume of the solution. Your lab instructor will ASSIGN you a SPECIFIC SOLUTION of sodium chloride (NaCl) to prepare, in terms of the PERCENT BY WEIGHT NaCl that the solution is to contain. By "percent by weight NaCl" is meant the weight of NaCl that would be contained in 100 g of solution. For example, a 35% by weight solution of NaCl would contain 35 g of NaCl per 100 g of solution. This solution could be prepared conceivably by adding 35 g of NaCl to 65 g of water (for a total solution weight of 100 g). Once the solution has been prepared, its density can be determined by WEIGHING a SPECIFIC VOLUME of the solution on the lab balance.

Summary

The density of some irregularly shaped chunks of rock is determined. The density of a solution of NaCl in water (of some assigned concentration) is determined. The composition of the solution made up must be within ± 1.5% of the assigned value.

Supplies

NaCl that has been kept stored in an oven to prevent moisture from caking the solid; chunks of rock. THE CHUNKS OF ROCK SHOULD BE RETURNED TO THE WASTE BEAKERS AFTER USE: DO NOT POUR THEM INTO THE SINKS!

CAUTION! WEAR SAFETY GLASSES AT ALL TIMES!!

CHEMICALS INFORMATION

Sodium chloride—No major health risks

Procedure

(1) Density of Rocks

Add water to a 25 mL graduated cylinder to about the 15 mL mark (the bottom of the curved water surface should line up with the 15 mL mark). Measure the volume to the nearest 0.1 mL and record this value in your lab notebook.

Weigh an empty 50 mL beaker on the lab balance.

Measure out approximately 20 g of the available rocks into the weighed beaker, and reweigh carefully (to the nearest 0.001 g).

Add the rocks to the water in the graduated cylinder, being careful to exclude air bubbles (which would change the volume).

Determine the water level after adding the rocks (again reading across the bottom of the curved water surface) to the nearest 0.1 mL. The CHANGE in water levels represents the volume occupied by the rocks.

DO NOT THROW THE ROCKS INTO THE SINKS OR THE WASTEBASKETS: A beaker is provided for collecting the rocks.

Calculate the density of the rocks, to **three significant figures**.

(2) Density of NaCl Solution

(a) Preparing the solution:

Based on the percentage assigned to you by your lab instructor, calculate the amounts of NaCl and water you will need to make 100 g of your solution.

Weigh an empty 150 mL beaker on the balance to the nearest 0.01 g. (Do not use a 250 mL beaker.)

Never weigh any chemical directly on the balance pan. Use either weighing paper or filter paper.

Carefully add the required amount of NaCl and reweigh (see note 1 below).

Carefully add the amount of water required and reweigh (see note 1 below).

Thoroughly dissolve the NaCl in the water by stirring for several minutes; the solution must be homogeneous before determining its density.

Note 1: In weighing out the NaCl and water, it is not necessary to weigh exactly the amounts calculated. Your solution's concentration may be within ± 1.5% of the assigned value: Quickly weigh as near as you can to the required amounts, but then use the actual amounts you weighed in calculating the exact percentage of your solution.

(b) Determining the density:

Weigh your empty, dry 25 mL graduated cylinder to the nearest 0.001 g.

Fill the graduated cylinder to exactly the 25 mL mark with a portion of your NaCl solution, and reweigh. THE DIFFERENCE IN WEIGHTS REPRESENTS THE MASS OF 25 mL OF YOUR SOLUTION.

Calculate the density of the solution, in g/mL, to **three significant figures**.

With your thermometer, determine the temperature of your solution. (The volume, and so the density, of a solution changes with temperature.) The temperature must be reported together with your density measurement.

Sodium chloride solutions can be poured down the drain.

REMEMBER TO TURN THE YELLOW COPY FROM YOUR LAB NOTEBOOK INTO YOUR TA.

Density Determinations
Report Sheet (40 points)

Name _____   Lab Day _____

Lab Instructor _____   Date _____

(Show calculations on the back of the page.)

1.   Density of rock chunks

Weight of rocks	Volume before adding rocks	Volume after adding rocks	Volume of rocks	Density of of rocks
_____	_____	_____	_____	_____

2.   Density of NaCl Solution   Assigned Concentration: _____ %

(a)   Preparation of Solution

Mass of NaCl needed   _____ g

Volume of distilled water needed   _____ mL

Weight of beaker   _____ g

Weight of beaker plus NaCl   _____ g

Weight of beaker plus NaCl plus water   _____ g

Actual % NaCl _____% (if different from assigned)

(b)   Density Determination

Weight empty graduate   _____ g

Weight of graduate plus solution   _____ g

Volume of solution   _____ mL

Density   _____ g/mL

Temperature of solution _____ °C

Density Determinations
Post-Lab Questions (30 points)

Name _____   Lab Day _____

Lab Instructor _____   Date _____

1.   A chunk of rock weighs 15.8 g and causes the water level in a graduated cylinder to rise from 22.3 to 32.5 mL. Calculate the density of the rock.

2.   Twelve grams of sodium chloride were dissolved in 52 mL (52 g) of distilled water, calculate the % sodium chloride in the solution.

3.   True or False: The density of a solution is dependent on the temperature of the solution.

4.   In this experiment, you were told to avoid getting air bubbles on the rock chunks while measuring your volume. Would air bubbles increase, decrease, or have no effect on the density of the rocks? Explain.

Density Determinations
Preliminary Questions (10 points)

Name _____   Lab Day _____

Lab Instructor _____   Date _____

(Show calculations on the back of the page.)

1. Ten grams of a sample of metal is added to 50.0 mL of water in a graduated cylinder. The final volume in the graduate is 52.8 mL. Calculate the density of the metal.

    _____g/mL

2. Twenty-five grams of NaCl is dissolved in 175.0 g of water. Calculate the percent by weight of NaCl in the solution.

    _____% NaCl

3. What is the proper disposal of your wet rocks?

4. A student is asked to make up a 16% sodium chloride solution. She makes the solution by weighing an empty beaker (which has a mass of 98.56 g). Sodium chloride is added to the beaker. The mass of the beaker and sodium chloride is 114.71 g. Lastly the student added 84 mL of distilled water to the beaker. Reweighing the beaker she finds that the mass of the beaker, sodium chloride, and distilled water is 196.14 g. What is the weight percentage of sodium chloride in this solution? (Use the back of this page to show your work, if needed).

Determination of a Simplest Formula

Introduction

An important use of the MOLE CONCEPT is in the determination of the SIMPLEST FORMULA of a compound. In this experiment, you will take a known mass of MAGNESIUM METAL and will convert it to MAGNESIUM OXIDE by heating in air. By subtracting the mass of metal initially taken from the mass of magnesium oxide present after the heating is completed, you can determine the mass of oxygen in the compound. From knowledge of the mass of magnesium initially taken, and the mass of oxygen gained from the air, it is a simple matter to find the molar ratio in which the elements combine. This gives us the simplest formula of the compound.

$$Mg(s) + \tfrac{1}{2} O_2(g) \rightarrow MgO(s)$$

Supplies

Crucible and cover; (CAUTION: The crucible and cover are FRAGILE) clay triangle; magnesium metal

Note

For meaningful results in this experiment, it is crucial that you WEIGH CORRECTLY, and with the correct PRECISION. The balances in our laboratories can weigh to the nearest THOUSANDTH of a gram (0.001 g). NO LESS PRECISION IS ACCEPTABLE!

CHEMICALS

Magnesium metal—Flammable solid

Procedure

CAUTION! IF MAGNESIUM METAL IS HEATED TOO STRONGLY IN AIR, IT WILL IGNITE PRODUCING AN INTENSELY BRIGHT FLAME (WHICH IS DAMAGING TO THE EYES). HEAT SLOWLY, WITH A SMALL FLAME, TO AVOID IGNITING THE MAGNESIUM, AND SPATTERING OF THE PRODUCTS. WEAR SAFETY GLASSES AT ALL TIMES!!

Obtain a crucible and cover. THE CRUCIBLE IS EXTREMELY FRAGILE: BE CAREFUL WITH IT. Wash it out with tap water and wipe with a towel to remove any loosely held solids.

From this point on, handle the crucible only with "crucible tongs" from your locker.

Set up a clay triangle that fits your crucible on a large metal ring on a ring stand. Heat the crucible and cover in the full heat of the burner flame for 5 minutes.

After heating, allow the crucible to cool for 5 minutes. Using the tongs, bring the crucible and cover to the balance and weigh (to nearest 0.001 g). Record the weight on the lab report page.

Add between 0.250 and 0.350 g of magnesium turnings to the crucible. Reweigh the crucible and contents. The weight of the crucible at this point, minus the weight of the empty crucible, gives the weight of magnesium used in the experiment.

Return the crucible to the clay triangle, and slowly begin to heat the crucible. TILT THE COVER of the crucible at an angle, so that the crucible's contents are slightly exposed to the air. If the contents begin to smoke greatly, REMOVE the heat and CLOSE THE COVER on the crucible (the "smoke" is the product of the reaction—magnesium oxide—which will be lost if you do not cover and stop heating).

Continue heating the magnesium until "smoke" no longer forms, then carefully remove the cover, and heat the crucible strongly for 3 to 4 minutes. The contents of the crucible should now be grayish/white, with perhaps a small amount of green.

Allow the crucible to COOL COMPLETELY to room temperature.

When the crucible is completely cool, add 5 to 6 drops of water. An additional product is magnesium nitride. Adding water converts the magnesium nitride to magnesium oxide and ammonia.

Begin heating the crucible slowly with a small flame—a small flame must be used to avoid splattering of the contents. After all the water has been evaporated from the crucible, heat the crucible with a full flame for 3 to 4 minutes. Allow the crucible to COOL COMPLETELY to room temperature.

When the crucible has cooled completely, weigh the crucible and cover to the nearest 0.001 g. The weight of the crucible now, after heating, minus the weight before heating, represents the weight of oxygen consumed by the magnesium in the reaction.

Dispose of your magnesium oxide product in the WASTE Magnesium Oxide bottle.

REMEMBER TO TURN IN THE YELLOW COPY OF YOUR LAB NOTEBOOK TO YOUR TA.

Calculations

1.   Calculate the number of moles of magnesium that you used in the experiment.

2.   Calculate the number of moles of oxygen that the magnesium consumed during the reaction.

3.   Calculate the RATIO of how many moles of oxygen are consumed per mole of magnesium used.

Given that the "correct" molar ratio for magnesium oxide is 1.00, calculate the "error" in your experiment:

$$\% \text{ Error} = \frac{(\text{your ratio - theoretical ratio})}{\text{theoretical ratio}} \times 100$$

Determination of a Simplest Formula
Report Sheet (40 Points)

Name _____   Lab Day _____

Instructor _____   Date _____

(a) Weight of empty crucible   _____

(b) Weight of crucible + magnesium   _____

(c) Weight of magnesium in crucible   _____

(d) Weight of crucible after heating   _____

(e) Weight of magnesium oxide present   _____

(f) Weight of oxygen absorbed by magnesium   _____

(g) Moles of magnesium used   _____

(h) Moles of oxygen absorbed   _____

(i) Ratio of magnesium/oxygen   _____

(j) % error in experiment   _____

Determination of a Simplest Formula
Post-Lab Questions (30 Points)

Name _____   Lab Day _____

Instructor _____   Date _____

1. An analysis shows that 2.97 g of iron metal combines with oxygen to form 4.25 g of an oxide of iron. What is the empirical formula of the compound?

2. An oxide of copper is decomposed forming copper metal and oxygen gas. A 0.500 g sample of this oxide is decomposed, forming 0.444 g of copper metal. What is the empirical formula of the gold oxide?

more questions on other side

3. Why should crucible tongs, and not fingers, always be used for handling the crucible and lid after the initial heating.?

4. When the magnesium is not allowed sufficient oxygen, some magnesium nitride, Mg_3N_2, forms.

   A. Write the balanced reaction that occurs when water is added to magnesium nitride.

   B. If the Mg_3N_2 is not decomposed, will the reported magnesium to oxygen ratio be high or low? Explain. Hint: The Mg:N mass ratio is 1:0.38 and the Mg:O mass ratio is 1:0.66.

Determination of a Simplest Formula
Preliminary Questions (10 Points)

Name _____   Lab Day _____

Instructor _____   Date _____

1.   Two grams of magnesium is heated as in this experiment, producing 3.33 g of magnesium oxide. Calculate:

   A.   The moles of magnesium present

   B.   The moles of oxygen present in the magnesium oxide

   C.   The molar ratio in which magnesium and oxygen have combined

2.   When the magnesium is not allowed sufficient oxygen, some magnesium nitride, Mg_3N_2, forms. How does one decompose any magnesium nitride?

CONSERVATION OF MASS

To a beginning student of chemistry one of the most fascinating aspects of the laboratory is the dazzling array of sights, sounds, odors, and textures that are encountered there. Among other things, we believe that this experiment will provide an interesting aesthetic experience. You will be asked to carry out a series of reactions involving the element copper and to carefully observe and record your observations. The sequence begins and ends with copper metal. Because no copper is added or removed between the initial and final steps, and because each reaction goes to completion, you should be able to quantitatively recover all of the copper you started with if you are careful and skillful. The following shows in an abbreviated form of the reactions of the cycle of copper:

$$Cu \rightarrow Cu(NO_3)_2 \rightarrow Cu(OH)_2 \rightarrow CuO \rightarrow CuSO_4 \rightarrow Cu$$

Like any good chemist, you will probably be curious to know the identity of each reaction product and the stoichiometry of the chemical reactions for each step of the cycle. They are listed below:

$$4HNO_3(aq) + Cu(s) \rightarrow Cu(NO_3)_2(aq) + 2H_2O(l) + 2NO_2(g)$$

$$Cu(NO_3)_2(aq) + 2NaOH(aq) \rightarrow Cu(OH)_2(s) + 2NaNO_3(aq)$$

Heat
$$Cu(OH)_2(s) \rightarrow CuO(s) + H_2O(l)$$

$$CuO(s) + H_2SO_4(aq) \rightarrow CuSO_4(aq) + H_2O(l)$$

$$CuSO_4(aq) + Zn(s) \rightarrow ZnSO_4(aq) + Cu(s)$$

These equations summarize the results of a large number of experiments but it is easy to lose sight of this if you just look at equations written on the paper. You can easily be overwhelmed by the vast amount of information found in an experiment and in chemistry textbooks. It is in fact a formidable task to attempt to learn or memorize isolated bits of information that are not reinforced by your personal experience. This is one reason why it is important to have a laboratory experience. Chemistry is preeminently an experimental science.

As you perform the experiment, watch closely and record what you see. It is easier to remember information that is organized by some conceptual framework. Observations and facts that have not been assimilated into some coherent scheme of interpretation are relatively useless.

Chemists look for relationships, trends, or patterns of regularity in organizing their observations of chemical reactions. The periodic table is a product of this kind of thinking. It groups the elements into chemical families. Each element bears a strong resemblance to other members of the same chemical family but also has its own unique identity and chemistry.

43

In a similar fashion, it is useful to classify reactions into different types. Several different kinds of classification schemes exist because no one scheme is able to accommodate all known reactions. A simple classification scheme we will use at the beginning is one based on ideas of combination, decomposition, and replacement. We present an outline and some examples of each type of reaction.

A SCHEME FOR CLASSIFYING CHEMICAL REACTIONS

Combination Reactions

Reactions that involve the combination of two or more pure substances to form a single substance.

Examples: $2Na(s) + Cl_2(g) \rightarrow 2NaCl(s)$

$CaO(s) + CO_2(g) \rightarrow CaCO_3(s)$

Decomposition Reactions

A single substance decomposes (often promoted by heat or light) into two or more different pure substance.

Examples: $2HgO(s) \xrightarrow{heat} 2Hg(s) + O_2(g)$

$NO_2(g) \xrightarrow{light} NO(g) + O(g)$

Single Replacement Reaction

A metal replaces another metal in a binary (made up of only two different elements) compound, or a nonmetal replaces another nonmetal in a binary compound.

Examples: $2AgNO_3(aq) + Cu(s) \rightarrow 2Ag(s) + Cu(NO_3)_2(aq)$

$MgCl_2(aq) + I_2(s) \rightarrow MgI_2(aq) + Cl_2(g)$

Double Replacement Reaction

Both partners in a binary compound exchange with a second binary compound.

Example: $AgNO_3(aq) + NaCl(aq) \rightarrow AgCl(s) + NaNO_3(aq)$

A second type of double replacement reaction is called a neutralization reaction. In this process an acid reacts with a base producing a salt and water.

Example: $HCl(aq) + NaOH(aq) \rightarrow NaCl(aq) + H_2O(l)$

As you carry out each step of the cycle of copper reactions, think about what is happening in each reaction and try to fit it into the classification scheme.

CHEMICAL RESPONSIBILITY

SAFETY ALERTS

1. Nitric acid, HNO_3, is toxic, corrosive, and an oxidant. It can cause severe skin burns. It also turns skin yellow on contact.

2. Sodium hydroxide, $NaOH$, is toxic and corrosive. It can cause severe skin burns and can cause severe damage to eyes. Wear your safety glasses.

3. Sulfuric acid, H_2SO_4, is toxic and corrosive. It can cause severe skin burns and can cause severe damage to eyes. Wear your safety glasses.

Some of the intermediate products of this reaction also are hazardous.

4. Copper(II) sulfate, $CuSO_4$, is toxic and an irritant.

5. Zinc sulfate, $ZnSO_4$, is an irritant and could be toxic.

6. Copper(II) nitrate, $Cu(NO_3)_2$, is toxic, corrosive and an oxidant.

    If you spill any of the above chemicals on your skin, immediately flush the affected area with water for several minutes. Quick action can prevent serious damage even from acids and sodium hydroxide.

7. Hydrogen gas, H_2, is a by-product of the reaction between zinc metal and sulfuric acid. It is a highly flammable gas. Make sure there are no open flames in the laboratory when performing this reaction. It is required for you to do this reaction in the hood to minimize any chances of a fire or explosion.

Chemical Disposal

1. All solutions should be poured into the Conservation of Mass waste bottle or 5-gallon container.

PROCEDURE

Caution!! Wear safety glasses at all times!!

I. **Conversion of Copper Metal to Copper(II) Nitrate**

1. Weigh a 0.250 to 0.350 g sample of copper metal (to the nearest 0.001 g) in a 150 mL beaker.

2. **Do the following step in the hood because of the generation of toxic nitrogen dioxide fumes (NO_2).**
    Add 10 to 15 mL of 6M nitric acid (HNO_3) to the beaker.

3. If the copper does not completely react, gently heat the beaker on a hot plate **in the hood** until all the copper metal has completely reacted.

    Caution! Nitrogen dioxide gas is generated in this step. It is toxic! Keep the reaction in the hood!

4. If more than half the liquid evaporates, add more 6M nitric acid. Do not allow the solution to boil.

NOTE: The blue solution is copper(II) nitrate; the brown gas evolved by this reaction is nitrogen dioxide, NO_2.

5. Complete the reaction listed on the report sheet.

II. **Conversion of Copper(II) Nitrate to Copper(II) Hydroxide**

6. Allow the beaker containing the copper(II) nitrate solution to cool to room temperature. **You can now take your beaker back to your work station.**

7. After the beaker has cooled, add 25 mL of distilled water to the beaker.

8. Pour about 10 mL of 6M NaOH into a 50 mL beaker.

Add 6M NaOH to the beaker until pH paper test blue (that is, the solution has become basic). The clear blue solution will turn to a milky green and then to a milky blue color. The solution will not test basic until it turns the milky blue color. Using a dropper add the NaOH by the dropper full until the solution turns green and then drop-wise until it turns blue.

Caution: Considerable heat may be given off in this step if any unreacted nitric acid remains in your reaction mixture. Adding the NaOH solution drop-wise will limit the spattering that may occur. Keep the beaker covered with a watch glass as much as possible during the addition of the first portions of the NaOH solution. Add the NaOH through the opening caused by the spout of the beaker.

9.   Test your solution with pH paper. It should be alkaline (basic) at this point. If the pH paper is green or blue, pH is 7 or greater. If it is not, add drop-wise additional NaOH until the solution tests basic.

NOTE: The pale blue precipitate, which is now present in your beaker is copper(II) hydroxide.

10.   Complete the reaction on the report sheet.

III.   Conversion of Copper(II) Hydroxide to Copper(II) Oxide

11.   Add 50 mL of distilled water to your pale blue copper(II) hydroxide solution.

12.   Boil the solution gentle with occasional stirring until the pale blue precipitate is converted to the black precipitate, copper(II) oxide (CuO). This will take at least 5 minutes of gentle boiling. Keep heating your solution until no blue color remains.

NOTE: This can be done on the hot plate in the hood or using your Bunsen burner at your lab desk.

13.   Complete the reaction on the report sheet.

IV.   Conversion of Copper(II) Oxide to Copper(II) Sulfate

14.   Allow your solution to cool to near room temperature without disturbing the beaker. This will allow your precipitate to settle to the bottom of your beaker. While your solution is cooling, prepare to perform a gravity filtration. Collect the filtrate (the liquid) in a 600 mL beaker to be used for waste.

15. After the precipitate has settled, pour off as much of the liquid portion as you can into the gravity funnel. When most of this liquid has filtered through, pour the remainder of the solution into the funnel. Rinse the beaker that contained the copper (II) oxide with about 5 mL of water and pour this into the funnel. When filtration is complete (or nearly complete), wash the brown-black solid on your filter paper with two 5 mL portions of distilled water (add the 5mL of water to the gravity funnel) and when most of the liquid has filtered through add the remaining 5 mL of distilled water.

16. Obtain about 15 mL of 3M sulfuric acid, H_2SO_4 (in a 50 mL beaker).

17. When most of the water has filtered, replace your waste beaker with the beaker that originally contained the copper(II) oxide. Pour the sulfuric acid onto the brown solid on the filter paper. When all the sulfuric acid has been added, gently stir the solution in the gravity funnel. Make sure not to put a hole in the filter paper. As the sulfuric acid reacts with the brown-black solid, it will turn to a clear blue solution. If any solid remains on the filter paper, obtain a clean 150 or 250 mL beaker and replace the one catching the blue solution with this clean one. Pour the blue solution back through the filtration apparatus. If any brown solid still remains, repeat this procedure until the filter paper is clean. At the end of this process rinse the empty beaker with about 5 mL of distilled water and transfer it to the beaker containing the blue solution.

NOTE: The solution now contains blue copper(II) sulfate solution.

18. Complete the reaction on the report sheet.

V.   Conversion of Copper(II) Sulfate to Copper Metal

19. Weigh out approximately 1 g of zinc metal (30 mesh).

20. Bring your beaker back to the hood. In the hood, slowly add the zinc to your copper(II) sulfate solution. If you add the zinc metal too quickly, it will react with excess sulfuric acid to produce a gas. This gas will bubble out of your solution, taking with it some of your liquid/solid mixture. This will result in a loss of copper metal. When the reaction has slowed down, you can bring the beaker back to your work station.

21. Allow the solution to stand until the blue color of copper(II) sulfate has completely disappeared. You should also notice the formation of copper metal at this point.

NOTE: The disappearance of the blue color is due to an electron transfer (that is, an oxidation-reduction) reaction taking place between metallic zinc and the copper(II) ions of copper(II) sulfate as follows:

$$Zn(s) + Cu^{2+}(aq) \rightarrow Zn^{2+}(aq) + Cu(s)$$
$$\quad\quad\quad blue \quad\quad\quad\quad colorless$$

22. Complete the reaction on the report sheet.

VI. Destruction of Excess Zinc Metal

NOTE: Once the blue color of the copper(II) ion has completely disappeared, it is safe to assume that all the dissolved copper (in the form of Cu^{2+}) has completely been reacted to form metallic copper. However, a closer look at the bottom of the beaker will also reveal the presence of unreacted zinc metal. This must be removed from our beaker before one can isolate the metallic copper. The metallic zinc is more reactive than metallic copper. The zinc metal can be converted to aqueous zinc sulfate solution by addition of sulfuric acid. H_2 gas is also a product of this reaction. The copper metal does not react with sulfuric acid.

23. Add about 10 to 15 mL of 3M sulfuric acid to your copper-zinc solution mixture.

24. Since it can take over an hour for all the zinc metal to react, you will stop at this point and finish this experiment next week.

25. From the supply table, obtain a marking pencil (red or blue, also called a china marker) and write your name or initials on the side of the beaker. This will allow you to identify your beaker next week. When the bubbles (hydrogen gas) are forming at a slow rate, obtain a piece of parafilm and cover your beaker (make sure you leave a small opening so excess gas can still escape). Leave your beaker on your benchtop, the staff in the chemistry stockroom will store your beakers for you until next week.

26. Complete the reaction on the report sheet.

VII. Washing the Copper Metal

27. Allow the metallic copper to settle to the bottom of your beaker undisturbed.

28. Pour off as much of the liquid as possible into your waste copper solutions beaker. If you fill this beaker use your 400 mL beaker for your waste solutions and continue pouring off as much liquid as possible. Be careful not to lose any copper.

NOTE: Your copper metal solution is still contaminated with dissolved ions, which would affect the weight of your copper if not removed.

29. Weigh a clean, dry evaporating dish to the nearest 0.001 g.

30. Add 10 mL of distilled water to the copper metal solution. Transfer the solution containing the copper metal to the evaporating dish. Use your polywash bottle filled with distilled water to make sure that all the copper has been removed from the beaker.

31. Pour off as much of the liquid as possible into your waste copper solutions beaker without losing any of your copper metal.

32. Add about 10 mL of distilled water to the evaporating dish, stir thoroughly and then allow the copper to settle to the bottom of the evaporating dish. Pour off the liquid into your waste beaker.

33. Repeat this process twice more, discarding the washings into your waste copper solutions beaker and exercising care not to allow any of the copper to be lost.

VIII. Recovering the Copper Metal

34. Obtain a 250 mL beaker from your lab draw. Half fill it with tap water. Prepare to heat the beaker (that is, put a wire gauze on a ring and the beaker on the wire gauze).

35. Place your evaporating dish on the top of the beaker.

36. Heat the apparatus with your Bunsen burner until the copper metal is dry. Occasionally remove the evaporating dish from the top of the beaker and tap the side of the dish on the desktop gently (**the evaporating dish will be hot: Use crucible tongs or paper towels to hold it**). If the copper moves freely, it is dry. If not, continue heating and repeat.

    NOTE: If the solid starts to turn a dark reddish-brown color, turn off the Bunsen burner. This indicates the formation of copper(II) oxide, CuO.

37. Allow the evaporating dish to cool to about room temperature. Dry the bottom of the evaporating dish and then weigh it.

38. Determine the mass of copper metal recovered.

REMEMBER TO TURN IN THE YELLOW COPY OF YOUR LAB NOTEBOOK TO YOUR TA.

Conservation of Mass
REPORT SHEET
(40 POINTS)

Name _____   Lab Day _____

Lab Instructor _____   Date _____

EXPERIMENTAL

Weight of copper metal initially taken: _____ g

Weight of cleaned evaporating dish: _____ g

Weight of copper metal and evaporating dish: _____ g

Weight of copper metal recovered: _____ g

% copper metal recovered: _____ g

For each step of the cycle, write the products of the reaction and balance the chemical equation. Using the classification scheme presented in this experiment, write the reaction type (combination, decomposition, single replacement or double replacement) for each reaction. Also record what you observe during each step in your lab notebook. Lastly answer any questions.

1. $Cu + HNO_3 \rightarrow$

What is in the solution after the reaction is complete?

Observations

2. $Cu(NO_3)_2 + NaOH \rightarrow$

Reaction type: _____

$HNO_3 + NaOH \rightarrow$

Reaction type: _____

What is formed in the solution besides $Cu(OH)_2$? _____

Observations:

3. $Cu(OH)_2 \xrightarrow{\text{heat}}$

Reaction type: _____

What is removed by the washing and decantation process at the end of step 12? (Consider the products of the reaction as well as reagents from previous steps.)

Observations:

4. $CuO + H_2SO_4 \rightarrow$

Reaction type: _____

What is in the solution at the end of this step? _____

Observations:

5. $Zn + CuSO_4 \rightarrow$

Reaction type: _____

$Zn + H_2SO_4 \rightarrow$

Reaction type: _____

What happens when the zinc is added? _____

What is removed by the washing and decantation at the end of this part of your experiment?

_____

Observations:

Conservation of Mass
Post-Lab QUESTIONS
(30 POINTS)

Name _____   Lab Day _____

Lab Instructor _____   Date _____

1.   A student performed this experiment, but was not particularly careful in following the procedure. Explain how each of the following procedural changes would affect his/her mass of copper metal recovered (too high, too low, or not changed)

   A.   Not all the copper metal had reacted with the nitric acid when the student started to add the sodium hydroxide.

   B.   Not enough sodium hydroxide was added, so the solution was still acidic when the beaker was heated to convert the copper(II) hydroxide to copper(II) oxide.

   C.   There was still some brown copper oxide present on the filter paper when the student moved on to the next step (adding zinc metal).

   D.   The copper metal was heated too long.

2.   A 0.325 g sample of copper was weighed out by a student to start this experiment.

   A.   How many moles of copper were weighed out?

   B.   How many moles of Cu^{2+} ions should be produced when the
          nitric acid was added to the copper metal?

   C.   When the sodium hydroxide was added to the solution, how many
          moles of $Cu(OH)_2$ should have formed?

   D.   The directions require you to add 1.00 g of zinc. If you assume a 100 %
          yield of copper, how many grams of zinc were added in excess?

   E.   If magnesium metal were used instead of zinc metal, what is the minimum mass, in
          grams, of magnesium metal that should be used to ensure that all of the copper ions
          in the solution is converted back to copper metal?

Conservation of Mass
Preliminary Questions
(10 POINTS)

Name _____ Lab Day _____

Lab Instructor _____ Date _____

1. Why is it necessary to perform the reaction of copper metal with nitric acid in the hood?

2. State the Law of Conservation of Mass.

3. Describe the hazards associated with following chemicals.

    A. Sodium hydroxide, NaOH

    B. Sulfuric acid, H_2SO_4

    C. Nitric acid, HNO_3

4. A student weighed out 0.385 g of copper metal to start the experiment. After completing the last step, the student had recovered 0.327 g of copper metal. What was the student's percent recovery?

Qualitative Inorganic Analysis-Identification of Six Solutions

Introduction

This experiment will give you a chance to be a sort of chemical Sherlock Holmes——ou will be given six bottles of UNKNOWN solutions, labeled only with a letter or number code, and you will IDENTIFY the solutions by their REACTIONS WITH EACH OTHER.

EVIDENCE for reaction when two solutions are mixed is based on observation of the following phenomena:

      (1) A precipitate (solid substance) is formed.
      (2) A gas is evolved from the solutions.
      (3) A color change is observed.
      (4) A characteristic odor is produced.

Each of these phenomena will be discussed briefly, with examples:

(1) A precipitate is formed.

Very often, when solutions of IONIC SOLUTES are mixed, one of the four possible COMBINATIONS of the ions involved will have a very SMALL SOLUBILITY in water, and a PRECIPITATE of that substance will form (imparting a CLOUDINESS to the solution until the substance settles out completely). For example, if 0.1 M solutions of BARIUM NITRATE and SODIUM CARBONATE are mixed (each of these substances is fairly water SOLUBLE), a PRECIPITATE instantly forms, since one of the four possible combinations of ions, BARIUM CARBONATE, is of very LOW SOLUBILITY.

$$Ba(NO_3)_2 + Na_2CO_3 \rightarrow 2NaNO_3 + BaCO_3 \qquad \text{(total reaction)}$$
$$Ba^{2+} + CO_3^{2-} \rightarrow BaCO_3 \qquad \text{(net ionic form)}$$

You can PREDICT which combinations of ions are likely to produce precipitates by consulting the "Solubility Rules" listed in your textbook in Chapter 4.

(2) A gas is evolved from the solutions.

Certain ions, such as CARBONATE and BICARBONATE, produce GASES when reacted with other ions (HYDROGEN ion, for example). Sometimes, the gas evolved can be IDENTIFIED by a characteristic odor [see also (4) below]; in other cases, the gas may have to be identified by other means.

$$Na_2CO_3 + 2HCl \rightarrow 2NaCl + H_2O + CO_2 \qquad \text{(total reaction)}$$

$$CO_3^{2-} + 2H^+ \rightarrow H_2O + CO_2 \qquad \text{(net ionic form)}$$

(3) A color change is observed.

The presence of certain ions can be confirmed by the addition of reagents that produce a COLOR CHANGE, which is CHARACTERISTIC of the ion under study; for example, AMMONIA, when added to a solution of COPPER(II) ION, produces a coordination complex of characteristic INTENSE BLUE color.

$$Cu^{2+} + 4NH_3 \rightarrow Cu(NH_3)_4^{2+}$$
$$\text{pale blue} \qquad\qquad \text{dark blue}$$

(4) A characteristic odor is produced.

Certain ions, especially when acidified, produce gases with CHARACTERISTIC ODORS (the solution may have to be heated to release the dissolved gas). As you will see in the experiment synthesizing sodium thiosulfate pentahydrate, when the thiosulfate ion is acidified and the solution heated, SULFUR DIOXIDE gas, with its characteristic choking odor, is released.

$$S_2O_3^{2-} + 2H^+ \rightarrow H_2S_2O_3 \rightarrow H_2O + S + SO_2$$

In this experiment, you will know WHAT the six solutions are, but you won't know WHICH BOTTLE contains which solution. Observations such as those described above for reactions AMONG the six solutions you are given can be related to the known chemistry of the given ions. Small portions (approx. 0.5 mL) of the six solutions are mixed, pairwise, and observations recorded and COMPARED to KNOWN reactions of the solutions involved. The six solutions provided to you are listed in the following table (see Table I); BEFORE COMING TO LAB, COMPLETE THE TABLE by writing in each box the expected product of the reaction. Indicate whether the products are solids, liquids, or gases (based on what was said above). If NO reaction is to be expected based on the substances to be mixed, write "N.R." in the box. You should also write net ionic reactions for the reactions in making your predictions

Summary

The identity of six solutions of ionized inorganic solutes is established by observing reactions between pairs of the solutions.

Supplies

0.5 M sodium carbonate; 0.1 M calcium chloride; 0.1 M barium nitrate; 0.05 M silver nitrate; 1.0 M hydrochloric acid; 1.0 M nitric acid (the solutions will be coded with letters); six small test tubes; six Pasteur pipets; pipet bulbs, 24 hole well plate.

CHEMICALS

Silver nitrate, AgNO$_3$—toxic and corrosive, may stain skin
Hydrochloric acid, HCl—toxic and corrosive, can cause skin burns
Nitric acid, HNO$_3$—corrosive and toxic
Barium nitrate, Ba(NO$_3$)$_2$—toxic by ingestion
Calcium chloride, CaCl$_2$—no major health risks
Sodium carbonate, Na$_2$CO$_3$—no major health risks

Procedure

CAUTION! WEAR SAFETY GLASSES AT ALL TIMES!!

Obtain a 24 hole well plate. Rinse it several times with distilled water to make sure it is clean.

Obtain about 2 mL (1/2 inch high) of each solution in its appropriate test tube.

Place a pipet in each test tube, and be careful not to switch pipets between solutions (thus contaminating them) during the following procedure.

Also obtain some rubber bulbs from the table in your lab (there may not be enough for each student to take six; if this is the case, move the bulb from pipet to pipet as needed).

Add about eight drops of the first solution to an empty well. Then add about eight drops of a second solution to this well. Observe what happens. Record your observations in your lab notebook. If no reaction occurs, swirl your well plate gently to mix the solutions. If still nothing has happened, write NR in your lab notebook for this combination.

Repeat this process until all possible combinations have been mixed.

A total of only 15 such tests is needed. (Note that mixing, e.g., solution A with solution B, is the same as mixing solution B with solution A.) You only need to complete Table II either above or below the diagonal; you can then complete the other half of the table by referring to the first half.

By comparing your results in Table II with the predictions made in Table I, you should be able to unambiguously identify each substance. Note in Table I that each substance reacts IN A UNIQUE WAY with the other five substances. (For example, sodium carbonate forms THREE PRECIPITATES and evolves GAS TWICE with the other reagents, and is the ONLY substance of the six to do this.)

REMEMBER TO TURN IN THE YELLOW COPY OF YOUR LAB NOTEBOOK TO YOUR TA.

DISPOSAL

Discard the contents of your well plate and test tubes in the beaker labeled **WASTE SOLUTIONS. Rinse your well plate and test tubes twice with a small amount of water and discard into the WASTE SOLUTIONS beaker.**

Dispose of the Pasteur pipets in the red plastic sharps-biohazard container. Return the pipet bulbs to the bag that they came from.

TABLE I (To be completed before coming to lab)

	Na_2CO_3	$CaCl_2$	$Ba(NO_3)_2$	$AgNO_3$	HCl	HNO_3
Na_2CO_3	XXXXX XXXXX					
$CaCl_2$		XXXXX XXXXX				
$Ba(NO_3)_2$			XXXXX XXXXX			
$AgNO_3$				XXXXX XXXXX		
HCl					XXXXX XXXXX	
HNO_3						XXXXX XXXXX

Copy the results onto the pre-lab page.

Experiment 11
Qualitative Inorganic Analysis
Identification of Six Solutions
Report Sheet (40 Points)

Name _____ Lab Day _____

Lab Instructor _____ Date _____

TABLE II (To be completed during lab period)

Solution Letters						
	XXXXX XXXXX					
		XXXXX XXXXX				
			XXXXX XXXXX			
				XXXXX XXXXX		
					XXXXX XXXXX	
						XXXXX XXXXX

Write NET IONIC EQUATIONS (there are seven) for all REACTIONS (precipitate formation and gas evolution) you have listed in Table I.

1.

2.

3.

4.

5.

6.

7.

Identification of Solutes

Code	Formula	Code	Formula
_____	_____	_____	_____
_____	_____	_____	_____
_____	_____	_____	_____

EXPERIMENT 11
Qualitative Inorganic Analysis—Identification of Six Solutions
Post-Lab Questions (30 Points)

Name _____   Date _____

Lab Instructor _____   Lab Day _____

1. One aspect of qualitative inorganic analysis involves the study of reactions between ions in solution. Evidence for reactions when two solutions are mixed is based on observations. Name two observations one might observe to indicate that a reaction has occurred when two solutions are mixed.

2. What are spectator ions?

3. What ions are formed when the following substances are dissolved in water?

    A. Na_2SO_4

    B. $CoCl_2$

    C.   Lithium carbonate

4. Complete the following reactions and then write the net ionic equation for each reaction.

    A. $Ba(NO_3)_2 + K_2SO_4 \rightarrow$

    B. $K_2CO_3 + HNO_3 \rightarrow$

5. An aqueous sample is known to contain Pb^{2+}, Cu^{2+}, or Na^+ ions. Treatment of the sample with both NaOH and LiCl solution produces a precipitate.

A. Which of the metal cations does the solution contain? Explain your reasoning.

B. Write all net ionic equations that could occur to justify your reasoning.

66

EXPERIMENT 11
Qualitative Inorganic Analysis—Identification of Six Solutions
Preliminary Questions (10 Points)

Name _____   Date _____

Lab Instructor _____   Lab Day _____

1. Fill in the following table with either the chemicals formed or no reaction upon mixing each pair of chemicals. (Copy over Table I below)

	Na_2CO_3	$CaCl_2$	$Ba(NO_3)_2$	$AgNO_3$	HCl	HNO_3
Na_2CO_3	XXXXX					
$CaCl_2$		XXXXX				
$Ba(NO_3)_2$			XXXXX			
$AgNO_3$				XXXXX		
HCl					XXXXX	
HNO_3						XXXXX

2. What is the proper disposal of all chemicals in this reaction?

3. Briefly explain any hazards associated with barium nitrate and silver nitrate.

4. A student is given the following labeled solutions: NaCl, $AgNO_3$, $FeCl_2$, HCl, and K_2CO_3. The data was collected and summarized in the following table. The student then receives the 5 test tubes labeled A, B, C, D, and E, which contain the five solution above. Data collected after mixing these solutions are given in the second table below.

Solution	K_2CO_3	$AgNO_3$	NaCl	$FeCl_2$	HCl
K_2CO_3	XXXXX XXXXX	Precipitate forms	No reaction	Precipitate forms	Gas forms
$AgNO_3$	Precipitate forms	XXXXX XXXXX	Precipitate forms	Precipitate forms	Precipitate forms
NaCl	No reaction	Precipitate forms	XXXXX XXXXX	No Reaction	No reaction
$FeCl_2$	Precipitate forms	Precipitate forms	No reaction	XXXXX XXXXX	No reaction
HCl	Gas forms	Precipitate forms	No reaction	No reaction	XXXXX XXXXX

Solution	A	B	C	D	E
A	XXXXX XXXXX	No reaction	Precipitate forms	Precipitate forms	No reaction
B	No reaction	XXXXX XXXXX	Gas forms	Precipitate forms	No reaction
C	Precipitate forms	Gas forms	XXXXX XXXXX	Precipitate forms	No reaction
D	Precipitate forms	Precipitate forms	Precipitate forms	XXXXX XXXXX	Precipitate forms
E	No reaction	No reaction	No reaction	Precipitate forms	XXXXX XXXXX

Using this data, identify each of the five solutions.

Solution A _____   Solution B _____

Solution C _____   Solution D _____

Solution E _____

Electrolytes

Introduction

ELECTROLYTES are substances that, when dissolved in water, permit PASSAGE OF AN ELECTRICAL CURRENT THROUGH THE WATER. Electrolytes produce IONS when dissolved in water, and the positive and negative ionic particles carry the electrical charge of the current through the water. Various substances can behave as electrolytes.

All SALTS are electrolytes, assuming they dissolve to a great enough extent in water. Salts in the solid state consist of an extended lattice of alternating + and - ions, and when added to water, this lattice breaks up, releasing the individual ions into the solution for conducting the current. One of the most important medical determinations is a patient's "electrolyte balance." Nerve impulses are sent through the body by electrical currents, and if the levels of electrolytes are incorrect, such nerve impulses may be incorrectly transmitted.

All ACIDS (H^+ producers) and BASES (OH^- producers) are electrolytes. Some acids, such as HCl, ionize completely when dissolved, and conduct electrical currents very well. Some bases, such as NaOH, also ionize completely when dissolved, and also conduct currents very well. Acids/bases such as HCl and NaOH, which conduct electrical currents very well, are called STRONG electrolytes. Other acids, such as acetic acid, do NOT ionize very well when dissolved in water, but rather exist in solution mostly as the molecular species. Some bases, such as ammonia and organic derivatives of ammonia, likewise do not ionize very well in water. Acids/bases that do NOT ionize well will not conduct very much electrical current, and are called WEAK electrolytes.

Some species NEVER ionize when dissolved in water, and will not conduct an electrical current through water at all. These are referred to as NON-electrolytes. Such substances are typically nonpolar, covalently bonded molecules (most typically, ORGANIC chemical substances).

The extent to which a solution will conduct an electrical current depends on other factors also. For example, we may have a substance such as NaCl, which is considered a strong electrolyte. But if a solution of NaCl is VERY DILUTE, there is so much DISTANCE between sodium ions and chloride ions, that effectively, the solution will not conduct electricity very well.

We can also change the ability of a solution to conduct electricity by chemically reacting the dissolved electrolyte. For example, we would expect a solution of acetic acid to conduct electricity very poorly. But if we were to add some chemical reagent to the solution that caused a chemical reaction to occur, then the electrolytic properties of the resulting solution may change.

CHEMICALS

Calcium chloride, $CaCl_2$—no major health risks

Sodium chloride, $NaCl$—no major health risks

Ammonium chloride, NH_4Cl—toxic by ingestion

Sodium sulfite, Na_2SO_3—irritant

Glucose or dextrose, $C_6H_{12}O_6$—no major health risks

Sucrose, $C_{12}H_{22}O_{11}$—no major health risks

Ethanol, C_2H_5OH—flammable liquid and toxic

Methanol, CH_3OH—flammable liquid and toxic

Hydrochloric acid, HCl—toxic, corrosive, and may cause burns

Acetic acid, $HC_2H_3O_2$—corrosive, toxic and may cause burns

Ammonia, NH_3—toxic, corrosive, may cause permanent fogging of soft contact lenses, and respiratory irritant

Sodium hydroxide, $NaOH$—corrosive, toxic and may cause burns

Procedure—METHOD 1 MACROSCALE

CAUTION! THE MEASUREMENT OF THE ELECTRICAL CONDUCTIVITY OF SOLUTIONS QUITE NATURALLY INVOLVES THE USE OF ELECTRICAL CURRENT. THERE IS THEREFORE A HAZARD OF ELECTRICAL SHOCK IN THIS EXPERIMENT. FOLLOW ALL INSTRUCTIONS PROVIDED BY YOUR INSTRUCTOR EXACTLY. DO NOT PERFORM ANY UNSPECIFIED MANIPULATIONS OF THE INSTRUMENTS INVOLVED. DO NOT TOUCH ANY PARTS OF THE APPARATUS THAT MAY CARRY THE ELECTRICAL CURRENT. WEAR SAFETY GLASSES AT ALL TIMES!!

In this experiment, you will study electrical conductivity of solutions by a "lightbulb" method:

The apparatus is connected to the 110 volt wall current (DANGER). The circuit contains a lightbulb, two electrodes for dipping into the solution to be tested, and certain internal resistances to match the current to the solutions. When the electrodes are dipped into the solution to be tested, the extent to which the lightbulb glows is an indication of the conductive properties of the solution. If the lightbulb glows brightly, the solution is a strong conductor. If the lightbulb glows only dimly, the solution is a poor conductor. If the lightbulb does not glow at all, then the solution is a non-conductor.

Your instructor will give particular instructions for the method. FOLLOW THESE INSTRUCTIONS EXACTLY TO PREVENT ANY DANGER OF ELECTRICAL SHOCK TO YOURSELF.

USE SMALL AMOUNTS OF ALL SOLIDS

1. Electrolytic properties of NaOH, HCl, Na$_2$SO$_3$, CaCl$_2$, NaCl, and NH$_4$Cl

A. Fill a 150 mL beaker about half way with water. Add about a quarter of a teaspoon of sodium chloride (NaCl) and stir to dissolve. Test for conductivity.

Repeat, using two of the other salts that are available in the lab.

Dispose of the salt solutions (NaCl, CaCl$_2$, Na$_2$SO$_3$ and NH$_4$Cl in the WASTE SALT SOLUTIONS bottle.

B. Obtain about 50 mL of 3 M HCl in a beaker and test for conductivity.

C. Obtain about 50 mL of 3 M NaOH in a beaker and test for conductivity.

Dispose of the HCl and NaOH solutions in WASTE ACID-BASE container.

2. Electrolytic properties of NH$_3$ and CH$_3$COOH

Retain both of these solutions for use in part 3.

A. Obtain about 50 mL of 3 M acetic acid in a beaker and test for conductivity.

B. Obtain about 50 mL of 3 M ammonia solution in a beaker and test for conductivity.

3. Reaction effects:

Add the ammonia solution from part 2 to the acetic acid solution from part 2. Test for conductivity.

The product of this neutralization reaction is the SALT ammonium acetate.

Dispose of this solution in the WASTE ACID-BASE container.

4. Electrolytic properties of isopropyl alcohol and sugar (sucrose or dextrose)

A. Fill a 150 mL beaker about halfway with water. Add about a teaspoon of sugar (sucrose or dextrose) and stir to dissolve. Test for conductivity.

B. Repeat using half a beaker of water and a few mL of alcohol as solute.

Dispose of both of these solutions in the WASTE SUGAR/ALCOHOL bottle.

5. Dilution effects:

A. Place about 40 mL of water in a 150 mL beaker. Add a "pinch" of NaCl and stir to dissolve. Test for conductivity.

B. Add 100 mL of water to the beaker to dilute the solution. Stir to dissolve and test for conductivity.

C. Transfer the contents of your 150 mL beaker to a 600 mL beaker.
Add another 100 mL water, stir, and test for conductivity.

D. Repeat the process of adding water and testing for conductivity until the beaker has been filled to capacity.

Dispose of this solution by pouring it down the drain.

Procedure—METHOD 2 MICROSCALE

ABOUT THE FLINN CONDUCTIVITY APPARATUS

This apparatus is safe since only a 9-volt battery is used. No electric shock is possible. Two (2) light emitting diodes (LED's) are used to add a quantitative plus to the meter. One LED is green; the second is red. The green LED requires more voltage than the red LED.

This meter cannot be easily shorted out since an on-off switch has been provided to disconnect the circuit even during storage. The electrodes are easily interchanged or replaced.

The meter is sufficiently sensitive that it displays even low conductivity. Tap water shows a low to medium conductivity.

Very low or no conductivity will result in neither LED lighting. Low conductivity will give a dim light to the red LED and while the green remains off. A scale is affixed to the meter and is also shown on the next page.

CONDUCTIVITY CHART

Scale	Red LED	Green LED	Conductivity
0	Off	Off	Low or None
1	Dim	Off	Low
2	Medium	Off	Medium
3	Bright	Dim	High
4	Very Bright	Medium	Very High

USE SMALL AMOUNTS OF ALL SOLIDS

1. Electrolytic properties of NaOH, HCl, Na_2SO_3, $CaCl_2$, NaCl, and NH_4Cl

A.   Half fill one well of a 24 hole well plate with distilled water. Add a small amount of NaCl to the well and stir with a stirring rod. Test for conductivity.

Repeat, using two of the other salts that are available in the lab

B.   Add about 20 drops of 1.0 M HCl (hydrochloric acid) to a well and test for conductivity. Repeat the procedure using 1.0 M NaOH.

2. Electrolytic properties of NH_3 and CH_3COOH

A.   Add about 20 drops of 1.0 M CH_3COOH (acetic acid) to a well and test for conductivity. Repeat this procedure with 1.0 M NH_3 (ammonia or ammonium hydroxide).

3. Reaction effects:

Add the 20 drops of the 1.0 M ammonia solution to the well containing the acetic acid solution. Test for conductivity.

The product of this neutralization reaction is the SALT ammonium acetate.

4. Electrolytic properties of isopropyl alcohol and sugar (sucrose or dextrose)

A.   Half fill a well with distilled water. Add a small amount of sucrose or glucose to the water. Stir with a stirring rod. Test for conductivity.

B.   Half fill a well with distilled water. Add 20 drops of isopropyl alcohol (rubbing alcohol) to the well. Mix and test for conductivity.

Dispose of the contents of your well plate into the beaker labeled HAZARDOUS WASTE, Electrolyte Solutions.

5. Dilution effects:

A.   Place about 40 mL of water in a 150 mL beaker. Add a "pinch" of NaCl and stir to dissolve. Test for conductivity.

B.   Add 100 mL of water to the beaker to dilute the solution. Stir to dissolve and test for conductivity.

C.   Transfer the contents of your 150 mL beaker to a 600 mL beaker.
Add another 100 mL water, stir, and test for conductivity.

D.   Repeat the process of adding water and testing for conductivity until the beaker has been filled to capacity.

Dispose of this solution by pouring it down the drain.

REMEMBER TO TURN IN THE YELLOW COPY OF YOUR LAB NOTEBOOK TO YOUR TA.

Electrolytes
Report Sheet (30 Points)

Name _____   Lab Day _____

Instructor _____   Date _____

Record your observations as to the electrical conductivities of the samples you tested: (strong electrolyte, weak electrolyte, or non-electrolyte)

NaCl_____

Na$_2$SO$_3$_____

CaCl$_2$_____

NH$_4$Cl _____

3 M HCl _____

3 M NaOH _____

3 M acetic acid _____

3 M ammonia _____

Upon mixing the acetic acid and ammonia_____

Sugar _____

Alcohol _____

Dilution Effects:

NaCl (before dilution) _____

After adding 100 mL H_2O _____

After adding 200 mL H_2O _____

After adding 300 mL H_2O _____

After adding 400 mL H_2O _____

After adding 500 mL H_2O _____

After adding 600 mL H_2O _____

Electrolytes
Post-Lab Questions (40 Points)

Name _____   Lab Day _____

Instructor _____   Date _____

1.   Although ammonia and acetic acid themselves are weak electrolytes, a mixture of these two solutions behaved as a strong electrolyte. Why?

2.   Write the reaction that occurs when ammonia and acetic acid are mixed.

3.   In one portion of this experiment you added a pinch of NaCl to a beaker, and you added progressively more and more water to the beaker. Explain why the light gets dimmer and dimmer.

4.   You are given three bottles labeled a, b, AND c. The bottles contain pentane (C$_5$H$_{12}$, a liquid), calcium chloride (CaCl$_2$) solution, and aqueous ammonium hydroxide (NH$_4$OH, ammonia dissolved in water). Describe a procedure to identify which liquid would be in each bottle.

5.   The device used in this experiment only measured the relative conductivity of the solution tested, that is, whether the solution contained a strong, weak or non-electrolyte. A more elaborate conductivity meter can determine a numerical value that can be related to the number of ions in solution. Arrange the following 0.1 M solutions in order of increasing conductivity and explain why you chose this order.

0.1 M FeCl$_3$ 0.1 M MgCl$_2$ 0.1 M NaCl   0.1 M glucose (C$_6$H$_{12}$O$_6$)

Electrolytes
Preliminary Questions (10 Points)

Name _____   Lab Day _____

Instructor _____   Date _____

1.   Classify each of the following as STRONG, WEAK, or NON-ELECTROLYTES:
      nitric acid, sulfuric acid, potassium hydroxide, calcium hydroxide, methylamine (an
      ammonia derivative), benzoic acid, salicylic acid, glucose, and oxygen gas

      Nitric acid   _____

      Sulfuric acid   _____

      Calcium hydroxide   _____

      Potassium hydroxide   _____

      Methylamine   _____
      (an ammonia derivative)

      Benzoic acid   _____

      Salicylic Acid   _____

      Glucose   _____

      Oxygen gas   _____

2.   Define the term weak electrolyte.

3.   What is the proper disposal of the salt solutions in this experiment?

Specific Heat

Introduction

This week in lab, and next week as well, you will be making measurements of HEAT energy. The basic unit of heat energy in the metric system is the JOULE. The "official" definition of the Joule is

$$1.0 \text{ Joule} = 1.0 \text{ kg m}^2 \text{ sec}^{-2}$$

However, this "official" definition of the Joule has little practical meaning. We will use a more understandable definition of the Joule:

4.18 Joules is the amount of heat energy required to change the temperature of 1.0 gram of water by 1.0 Celsius degree.

This definition is usually written in a different manner: We are talking about the substance WATER in this definition, and are indicating the amount of heat required to change a UNIT AMOUNT of water (1.0 gram) by a SINGLE DEGREE Celsius. The quantity of heat required to change the temperature of a unit amount of substance by one degree Celsius is called the SPECIFIC HEAT of the substance, S. So we can say that 4.18 Joules per gram per degree is the SPECIFIC HEAT of water.

$$S_{water} = 4.18 \text{ Joules/g}^o\text{C}$$

Water is a substance that is used very often in the measurement of TRANSFERS of heat energy. A MEASURED AMOUNT of water is used to COLLECT the heat energy being emitted by a warmer substance or by a chemical reaction. By determining the TEMPERATURE CHANGE undergone by a MEASURED SAMPLE of water as it absorbs heat energy from a source, it is very simple to calculate the quantity of heat energy ABSORBED BY THE WATER.

Suppose we have a reservoir containing 150 g of water at 22°C. Suppose the temperature of the water is somehow increased to 38°C.

The water must have ABSORBED HEAT ENERGY for its temperature to have increased. The QUANTITY OF HEAT ENERGY, in JOULES, would be calculated using the SPECIFIC HEAT OF WATER, the MASS OF WATER in grams, and the TEMPERATURE CHANGE undergone by the water:

HEAT = (Specific heat) x (Mass) x (Temperature change)

HEAT = (4.18 Joule / g°C)(150 g)(38-22°C) = 10032 Joules

Oftentimes, the amount of heat involved in a simple process will be a LARGE NUMBER of Joules. For this reason, heat flows are usually given in KILOJOULES, rather than Joules:

> 1000 Joules = 1.000 kilojoule
> 10032 Joules = 10.032 kilojoules

The concept of "specific heat" is applicable to ANY SUBSTANCE: The specific heat is the number of Joules of heat energy required to raise the temperature of 1.0 gram of the given substance by 1.0 Celsius degree.

One simple way the specific heat of a substance can be measured is used in this experiment. The method makes use of the CONSERVATION OF ENERGY principle, that energy cannot be "lost" during a process. The method is best illustrated by an example:

Suppose we have a piece of "substance X" weighing 50 g, and we heat the piece of "substance X" to a particular temperature, for example, $100^{o}C$. Suppose we have an insulated reservoir containing 150 g of water at $22^{o}C$. Suppose we transfer the 50 g of "substance X" at $100^{o}C$ to the reservoir of water.

Obviously, when the hot piece of "substance X" is transferred to the cool water, a transfer of heat energy will take place. The hot piece of "substance X" will cool down, and the cool water will warm up. The piece of "substance X" will transfer heat energy to the water, until both substances reach the SAME TEMPERATURE. For example, suppose the system finally came to equilibrium at $38^{o}C$.

According to the conservation of energy principle, the quantity of heat LOST BY THE HOT OBJECT will exactly EQUAL the quantity of heat GAINED BY THE WATER.

The amount of heat GAINED BY THE WATER is given by:

> $(4.18$ Joules $/ g^{o}C)(150$ g$)(38-22^{o}C) = 10032$ Joules

The amount of heat LOST BY SUBSTANCE X is given by

> $(S_{substance X})(50$ g$)(100-38^{o}C)$

Since the amount of heat LOST by substance X must EQUAL the amount of heat GAINED by the water, then the sample of substance X must have LOST 10032 Joules. From this, we can solve for the SPECIFIC HEAT of substance X:

$(S_{substance X})(50$ g$)(100-38^{o}C) = 10032$ Joules

$S_{substance X} = 10032$ Joules $/ (50g)(100-38^{o}C) = 3.24$ Joule/$g^{o}C$

In this experiment, you will use the method outlined above to determine the specific heat of two substances: A metal and glass.

Procedure

CAUTION! AVOID SPLASHING OF THE BOILING WATER BATH USED BELOW. WEAR SAFETY GLASSES AT ALL TIMES!!

Part 1:   Determination of the Specific Heat of a Metal

Set up a ring stand, ring, and wire gauze, and begin to heat a 600 mL beaker about 3/4 full of water. Bring the water to the boiling point (100°C).

Clean out your large test tube (25 x 200 mm) and dry it out with a paper towel.

Weigh an empty clean, dry 150 mL beaker. Weigh 25.0 g of metallic copper into the beaker.

Obtain a plastic foam coffee cup. This will serve as the water reservoir in which temperature changes will be measured. An insulated container such as the coffee cup is called a CALORIMETER, and prevents any heat from being lost to the room during the measurements to be performed.

With a graduated cylinder, measure 50.0 mL (= 50.0 g) of water into the coffee cup.

By now the water in the beaker should be boiling smoothly. Suspend the large test tube in the beaker of boiling water using a clamp. Transfer the 25.0 g of metal to the test tube. By this method, the metal in the test tube will be heated to the temperature of the boiling water bath.

Boil the metal in the test tube in the water bath for at least 10 minutes, to ensure that the metal will have reached the temperature of the water bath.

After the metal has been heating for 10 minutes, use your thermometer to determine the exact temperature of the water in the coffee cup. Record this temperature to the nearest tenth of a degree.

Pour the heated metal into the water in the coffee cup. With your thermometer, gently stir the contents of the coffee cup, watching the temperature rise of the water on the thermometer scale. When the temperature has stopped rising (as heat is transferred from the hot metal to the cool water), record the highest temperature reached by the system.

From the MASSES of metal and water used, and from the TEMPERATURE CHANGES undergone by the metal and water, calculate the SPECIFIC HEAT of the metallic substance.

Obtain the data from two of your neighbors. Use all three sets of data to find an average value for the specific heat of copper.

PURE METALLIC SUBSTANCES ARE VERY EXPENSIVE. A BEAKER IS PROVIDED FOR COLLECTING THE METAL YOU HAVE USED. DO NOT THROW THE METAL AWAY.

Part 2: Determination of the Specific Heat of Glass

Repeat the procedure used above, only substitute 25.0 g of glass beads for the 25.0 g of metallic copper. Determine the specific heat of glass from your data.

Obtain the data from two of your neighbors. Use all three sets of data to find an average value for the specific heat of the glass beads.

A BEAKER IS PROVIDED FOR COLLECTING THE GLASS BEADS AFTER YOU ARE THROUGH USING THEM. DO NOT THROW THEM AWAY OR SPILL THEM ON THE FLOOR OR IN THE SINKS.

Return the styrofoam cup to the plastic bin on the supply table.

Specific Heat
Report Sheet (30 Points)

Name _____   Date _____

Lab Instructor _____   Lab Day _____

Part 1: Specific Heat of Copper Metal

	Your Data	Neighbor 1	Neighbor 2
Weight of copper taken, grams	_____	_____	_____
Weight of water taken, grams	_____	_____	_____
Temperature of heated copper, ^{o}C	_____	_____	_____
Temperature of cool water, ^{o}C	_____	_____	_____
Final temperature reached, ^{o}C	_____	_____	_____
Heat GAINED by water, Joules	_____	_____	_____
Heat LOST by copper, Joules	_____	_____	_____
Specific heat of copper, Joules/g^{o}C	_____	_____	_____

Average value of specific heat of copper metal _____

Part 2: Specific Heat of Glass Beads

	Your Data	Neighbor 1	Neighbor 2
Weight of glass taken, grams	_____	_____	_____
Weight of water taken, grams	_____	_____	_____
Temperature of heated glass, $^{\circ}$C	_____	_____	_____
Temperature of cool water, $^{\circ}$C	_____	_____	_____
Final temperature reached, $^{\circ}$C	_____	_____	_____
Heat GAINED by water, Joules	_____	_____	_____
Heat LOST by glass, Joules	_____	_____	_____
Specific heat of glass, Joules/g$^{\circ}$C	_____	_____	_____

Average value of specific heat of glass beads   _____

Specific Heat
Post-Lab Questions (40 Points)

Name _____   Date _____

Lab Instructor _____   Lab Day _____

1.   A solar panel is used to heat water in a tank. How much heat energy is needed to raise the temperature of 1000 L of water from an initial temperature of 23.0°C to a final temperature of 65.0°C.

2.   Determine the specific heat of an unknown metal when 25.0 g of the metal, initially at 80.0°C is added to 100.0 mL of water initially at 20.0°C. The final temperature of the mixture is 21.6°C.

more questions on the other side

3.　Thirty grams of silicon at 80.0°C is added to 70.0 mL of water at 20.0°C. Calculate the final temperature after the two substances are mixed together. The specific heat of water is 4.184 J/g°C and the specific heat of silicon is 0.7121 J/g°C.

4.　A cube of brass, 2 cm on a side, is heated until the temperature of the brass is 75°C. The cube is quickly added to 100 mL of water at 23°C. What is the final temperature of the mixture? The specific heat of brass is 0.385 J/g°C. The density of brass is 8.45 g/cm^3.

Specific Heat
Preliminary Questions (10 Points)

Name _____ Date _____

Lab Instructor _____ Lab Day _____

Preliminary Questions

1. The specific heat of copper is 0.385 J/g°C. Calculate the final temperature when 25.0 g of copper metal at 100°C is added to 50 mL of water at 20°C.

2. If 75.0 grams of water is heated from 32.6°C to 78.9°C, how many kilojoules of heat does the water absorb?

more questions on other side

3.   A 25.0 g sample of zinc metal at 85.0°C is added to 75.0 g of water initially at 18.0°C. The final temperature is 20.0°C.

A.   How much heat is gained by the water?

B.   How much heat is lost by the zinc metal?

C.   From the data in this problem, calculate the specific heat of zinc metal.

Heat of Reaction

Introduction

HEAT is defined as "energy transferred from one system to another because of a difference in temperature." Heat always flows FROM the HOTTER to the COLDER system. In this experiment, you will determine HEATS OF REACTIONS carried out in DILUTE SOLUTION in an insulated container—a simple CALORIMETER. In such cases, if the reaction is EXOTHERMIC, heat flows FROM the reacting substances TO the solution, thereby RAISING ITS TEMPERATURE. If the reaction is ENDOTHERMIC, heat flows FROM the solution TO the reacting substances, thereby LOWERING THE TEMPERATURE of the solution.

The UNIT OF HEAT based on experimental measurements is the CALORIE, which is defined as the amount of heat required to change the temperature of 1.0 g of WATER by 1.0°C at 15°C. The correct unit of heat in SI units is the JOULE, however. Formally, the Joule is defined as 1.0 Kg m^2/sec^2; also, 1.0 calorie = 4.184 Joules. The kilojoule (kJ) is 1000 Joules.

The SPECIFIC HEAT of a substance is the AMOUNT OF HEAT REQUIRED TO RAISE THE TEMPERATURE of 1.0 g of the substance by 1.0°C. For example, the SPECIFIC HEAT OF WATER would be 1.00 cal/g°C by this definition. In terms of Joules, the specific heat of water would be 4.184 J/g°C. Specific heats of substances vary somewhat with temperature. Over the temperature range of this experiment, however, you may assume that the specific heats remain constant. Also, the specific heat of a solution is NOT generally the same as the specific heat of the pure solvent. In this experiment, you may also assume that the specific heats of the solutions are very near to that of pure water (since the solutions are relatively dilute).

All chemical reactions either EVOLVE (give off) or ABSORB (take in) heat. The HEAT OF REACTION is the number of kilojoules evolved or absorbed PER MOLE of reactant or product, based on the equation for the reaction as written. When reactions are conducted at CONSTANT PRESSURE (for example, in the lab under the pressure of the atmosphere), HEAT OF REACTION is called ENTHALPY OF REACTION and is given the symbol ΔH. There are some conventions concerning the mathematical sign of ΔH:

(a)   In EXOTHERMIC reactions, heat is evolved by the reacting system; kilojoules are LOST by the system and are given a NEGATIVE sign. For example:

$$H_2(g) + \tfrac{1}{2} O_2(g) \rightarrow H_2O(l) \quad \Delta H = -285.8 \text{ kJ/mole of } H_2O$$

(b)   In ENDOTHERMIC reactions, heat is absorbed by the reacting system; kilojoules are GAINED by the system and are given a POSITIVE sign. For example:

$$\tfrac{1}{2} H_2(g) + \tfrac{1}{2} I_2(g) \rightarrow HI(g) \quad \Delta H = +25.9 \text{ kJ/mole of HI}$$

Note: If either of the above reactions are written in REVERSE, products become reactants, "exo" becomes "endo" (and vice versa), and the sign of ΔH must be changed. For example:

$$HI(g) \rightarrow \tfrac{1}{2} H_2(g) + \tfrac{1}{2} I_2(g) \qquad \Delta H = -25.9 \text{ kJ/mole of HI}$$

Since ΔH for a chemical reaction depends only on the initial and final states of the system, it is said to be INDEPENDENT of the path of reaction taken. That is, ΔH is the SAME for a given reaction, whether it is carried out in ONE STEP OR IN SEVERAL STEPS. This property of ΔH is referred to as HESS's LAW, and is a consequence of the LAW OF CONSERVATION OF ENERGY.

Let us first consider a "general" example of the application of Hess's Law: Compound A reacts to form Compound B, but also can react to form Compound C.

$$(1) \ A \rightarrow B \quad \Delta H = -x \text{ kJ/mole}$$

$$(2) \ A \rightarrow C \quad \Delta H = -y \text{ kJ/mole}$$

From the preceding data, ΔH for the reaction of Compound B to form Compound C can be calculated:

$$A \rightarrow C \quad \Delta H = -y \text{ kJ/mole}$$

$$B \rightarrow A \quad \Delta H = +x \text{ kJ/mole} \quad \text{(note sign change on reversing equation)}$$

$$\overline{}$$

$$B \rightarrow C \quad \Delta H = (x - y) \text{ kJ/mole} \qquad \text{by addition}$$

Note that in the solution of the problem, reaction (1) above was written in REVERSE, and so the SIGN of ΔH was CHANGED. In the application of Hess's law, equations can be added together, reversed, doubled, halved, subtracted, etc.—CONCURRENTLY PERFORMING THE SAME ALGEBRAIC MANIPULATION ON THE SIGN AND MAGNITUDE OF ΔH. For example, if the quantities in a chemical reaction are doubled, ΔH should be multiplied by two.

Now let's consider a specific problem. The heat of combustion of CARBON burning to form CARBON DIOXIDE, and the heat of combustion of CARBON MONOXIDE to form CARBON DIOXIDE can be determined by CALORIMETRY:

$$C(s) + O_2(g) \rightarrow CO_2(g) \qquad \Delta H = -393 \text{ kJ/mole of } CO_2$$

$$CO(g) + \tfrac{1}{2} O_2(g) \rightarrow CO_2(g) \qquad \Delta H = -283 \text{ kJ/mole of } CO_2$$

What we would like to calculate is ΔH per mole of CO formed for the reaction of CARBON and MOLECULAR OXYGEN to give carbon monoxide. Based on what was said above in the general example, you should be able to figure out ΔH for the process, which is -110 kJ/mole of CO.

You will note that in thermochemical equations the PHYSICAL STATES of reactants and products (solid, liquid, gaseous, or aqueous solution) are given. This is ESSENTIAL because CHANGES OF STATE INVOLVE HEAT CHANGES: for example, heat of melting, heat of vaporization, heat of hydration, etc. A given reaction producing water in the VAPOR phase will produce LESS HEAT than the same reaction producing LIQUID water. The DIFFERENCE represents the heat of VAPORIZATION (or condensation) of water.

Today's experiment involves determining the ΔHs for reactions in dilute aqueous solution. Thus, heats of hydration of ions produced are included in the overall heats of reaction measured. The reactions are:

(1) $NaOH + HCl \rightarrow NaCl + H_2O$

(2) $Mg + 2HCl \rightarrow MgCl_2 + H_2$

These reactions are better written as "net ionic equations" eliminating "spectator ions" in each case:

(1a) $OH^-(aq) + H^+(aq) \rightarrow H_2O(l)$

(2a) $Mg(s) + 2H^+(aq) \rightarrow Mg^{2+}(aq) + H_2(g)$

Having determined ΔH for each of the above reactions, you are to calculate ΔH for the following reaction:

(3a) $Mg(s) + 2H_2O(l) \rightarrow Mg^{2+}(aq) + 2OH^-(aq) + H_2(g)$

The following example is designed to aid you in your calculation of ΔHs from temperature change data in your experiment:

When 150 mL of 0.40 M HCl is neutralized with 165 mL of 0.40 M ammonia solution, the temperature is observed to rise by 2.36°C for this process. Assuming the specific heats of the dilute solutions to be 4.184 J/g°C, and the densities of the solutions to be very nearly 1.00 g/mL, calculate the heat of reaction.

Since Cl⁻ is a spectator ion in this process, the net ionic equation is just

(4a) $H^+(aq) + NH_3(aq) \rightarrow NH_4^+(aq)$

To solve the problem, first calculate the heat evolved by the reaction:

$(315 \text{ g})(4.184 \text{ J/g°C})(2.36°C) = 3110 \text{ J evolved}$

Note: 150 mL plus 165 mL gives a total of 315 mL. Assuming that the density of this solution is 1.00 g/mL, the mass of the solution is 315 g.

Next, calculate the number of Joules evolved PER MOLE of HCl reacted. (The slight excess of NH_3 remains unreacted.)

Note: mole = M x L

$$(0.15 \text{ liters})(0.40 \text{ moles/liter}) = 0.060 \text{ moles HCl}$$

$$(3110 \text{ J}) / (0.060 \text{ moles}) = 51840 \text{ J/mole} = 51.84 \text{ kJ/mole}$$

Since the reaction is EXOTHERMIC (the temperature increased on reaction),

$$\Delta H = -51.84 \text{ kJ/mole for this process.}$$

Summary

Molar heats of reaction are determined for the following reactions:

(1) $H^+(aq) + OH^-(aq) \rightarrow H_2O(l)$

(2) $Mg(s) + 2H^+(aq) \rightarrow Mg^{2+}(aq) + H_2(g)$

Based on the data obtained, the heat of reaction is calculated for the reaction

(3) $Mg(s) + 2H_2O(l) \rightarrow Mg^{2+}(aq) + 2OH^-(aq) + H_2(g)$

Supplies

2 M HCl; 2 M NaOH; plastic foam cup (the calorimeter); magnesium metal.

CHEMICALS

HCl (hydrochloric acid)—toxic and corrosive
NaOH (sodium hydroxide)—toxic, corrosive and easily absorbed through skin
For these substances, prevent contact with skin and eyes. If any of these substances get on your skin, wash the contaminated area immediately with plenty of cold water.
Magnesium metal—flammable solid

Procedure

CAUTION! WEAR SAFETY GLASSES AT ALL TIMES!

(1)   ΔH for reaction of HCl and NaOH solution

Add 75 mL of 2.0 M HCl solution to the styrofoam cup, which will serve as a simple calorimeter.

Rinse and dry the graduated cylinder and measure out 75 mL of 2.0 M NaOH solution into it.

Measure and record the temperatures of both solutions TO THE NEAREST 0.2°C. Wipe the thermometer dry between measurements to avoid any mixing of acid with base. The two temperatures should be within 0.5°C of each other before continuing (if they are not, warm or cool the solutions until the temperatures are the same).

Pour all the NaOH solution quickly into the HCl solution.

Stir the reaction mixture with the thermometer and during the stirring, watch the temperature change. Record TO THE NEAREST 0.2°C the HIGHEST TEMPERATURE reached.

Discard your solution into the WASTE ACID-BASE CONTAINER.

Use your data to calculate ΔH per mole of water formed in the acid-base reaction.

For your calculation, assume that the total weight of the solution is 150 g, and take the average of the two starting temperatures as your initial temperature.

(2)   ΔH for the reaction of Mg metal with HCl solution

Rinse the styrofoam cup with water and dry it. Prepare 150 mL of 1.0 M HCl by appropriate dilution of 2.0 M HCl solution. Add the diluted solution to the styrofoam cup.

Weigh a 0.300 g (not 3.00 g) sample of magnesium metal on a clean, dry watch glass. Determine the temperature of the 1.0 M HCl solution to the nearest 0.2°C.

Add the magnesium to the HCl in the cup all at once, stirring the reaction mixture constantly with the thermometer. Record the HIGHEST temperature reached on the complete reaction of the magnesium.

Discard your solution into the WASTE ACID-BASE CONTAINER.

Based on your data, calculate ΔH per mole of Mg reacted. For your calculations, assume that the total weight of the solution is 150 g.

(3) ΔH for the reaction of Mg metal with water

Using data you obtained in parts (1) and (2) above, calculate ΔH for the reaction between magnesium metal and water (per mole of Mg reacted).

In all calculations above, assume the specific heat of the solutions is 4.184 J/g°C. Express your ΔHs in kJ/mole of the indicated reactant. Also, be sure you indicate the sign (+ or -) of ΔH correctly (remember that a temperature rise means an exothermic process).

REMEMBER TO TURN IN THE YELLOW COPY OF YOUR LAB NOTEBOOK TO YOUR TA.

Heat of Reaction
Report Sheet (30 Points)

Name _____   Lab Day _____

Lab Instructor_____   Date _____

Part 1

INITIAL temperatures, °C:

HCl solution _____   NaOH solution _____

FINAL Temperature, °C   _____

Temperature CHANGE, °C   _____

Heat evolved   _____ Joules

_____ kilojoules

Moles of water (from the balanced chemical equations) _____

ΔH per mole of water formed, kilojoule/mole _____

Part 2

Weight of magnesium   _____ grams

Moles of magnesium   _____ moles

INITIAL temperature, °C   _____

FINAL temperature, °C   _____

Temperature change, °C   _____

Heat evolved   _____ Joules

_____ kilojoules

ΔH per mole of Mg reacted   _____ kilojoule/mole

Part 3

For the reaction of Mg metal with excess water (from Hess's Law calculation)

ΔH per mole of magnesium reacted _____ kilojoule/mole

Show all calculations below:

Heat of Reaction
Post-Lab Questions (40 Points)

Name _____   Lab Day _____

Lab Instructor_____   Date _____

1.   True or False: Heat always flows from the hotter system to the cooler system.

2.   Were the reactions you performed exothermic or endothermic? Why?

3.   Define specific heat.

4.   True or False: Enthalpy is defined as the change in heat at constant pressure

5.   True or False: The amount of heat absorbed or evolved is proportional to the masses of the reacting materials.

More questions on other side

6. Consider the reaction

    $HCl + NaOH \rightarrow NaCl + H_2O$

    which you also performed in the experiment Stoichiometry of Two Reactions. Given the information below:

    Initial temperature °C _____

    HCl solution __22__   NaOH solution __22__

    Final temperature °C __26.1__

    Temperature Change °C _____

A. Calculate the amount of heat evolved when 15 mL of 1.0 M HCl was mixed with 35 mL of 1.0 M NaOH.

B. Calculate $\Delta H°$ per mole of water formed from this data:

    Heat evolved   _____ Joules

                                    _____ kilojoules/mole

Heat of Reaction
Preliminary Questions (10 Points)

Name _____   Lab Day _____

Lab Instructor_____   Date _____

1.   Define "heat of reaction"

2.   If 150 grams of water changes temperature by 7.2°C, how much heat energy flows?

3.   Describe how you would make 150 mL of a 1.0 M HCl solution starting with 2.0 M HCl.

4.   Use Hess's Law to calculate the heat of formation of $CO(g)$ given the following information:

$C(s) + O_2(g) \rightarrow CO_2(g)$ $\Delta H° = -393.5$ kJ

$CO(g) + \frac{1}{2} O_2(g) \rightarrow CO_2(g)$ $\Delta H° = -283.5$ kJ

Find $\Delta H°$ for the reaction: $C(s) + \frac{1}{2} O_2(g) \rightarrow CO(g)$

Atomic Emission Spectra—Flame Tests

Introduction

In the simple Bohr atomic theory you have discussed in lecture, when a HYDROGEN ATOM is bombarded with sufficiently HIGH ENERGY, the electron of the hydrogen atom ABSORBS some of the energy and MOVES TO A HIGHER ORBIT (an orbit that is farther from the nucleus). At a later time, the electron DROPS BACK TO ITS ORIGINAL ORBIT and RELEASES the "extra" energy it had absorbed. In some instances, the energy released by an electron in dropping back to a lower orbit is of such a WAVELENGTH that it is visible to the eye as COLORED LIGHT.

Since the orbits in the hydrogen atom are FIXED DISTANCES apart, corresponding to DISCRETE DIFFERENCES IN ENERGY, the light emitted by hydrogen atoms when excited always occurs at EXACTLY THE SAME wavelengths (colors). The hydrogen atom emits visible light at the following wavelengths (given in nanometers, with the corresponding color listed): 410 (violet), 434(blue), 486 (green), and 656 (red). In addition to the light that is visible to the pigments of the eye, other wavelengths of radiation are also emitted by the hydrogen atom (e.g., ultraviolet, infrared). These emissions, however, also occur at characteristic wavelengths.

The idea of an excited atom emitting light applies in general, in particular with the metallic elements (found at the left and bottom of the Periodic Table of the elements). In most cases, the WAVELENGTHS (colors) of light emitted by excited atoms are so CHARACTERISTIC and SPECIFIC that they can be used to IDENTIFY the atoms. For example, when table salt (NaCl) is spilled into a flame, an intense bright ORANGE color results: This orange color is due to a PARTICULAR EMISSION WAVELENGTH that is CHARACTERISTIC of sodium atoms. If such an orange color is observed in the light emitted by an "unknown" excited atom, the unknown atom is almost certainly sodium. The study of the characteristic wavelengths emitted by excited atoms is called EMISSION SPECTROSCOPY.

In today's experiment, you will observe the colors emitted by certain KNOWN elements when they are excited in the Bunsen burner flame. You will then be given an UNKNOWN SAMPLE containing one of the elements previously studied, and will attempt to identify the unknown by the color of the flame produced when excited.

CHEMICALS

NaCl (sodium chloride)—irritant
KCl (potassium chloride)—no health related risks
$SrCl_2$ (strontium chloride)—no health related risks
$BaCl_2$ (barium chloride)—toxic by ingestion
$CaCl_2$ (calcium chloride)—no health related risks
$CuCl_2$ (copper(II) chloride—toxic
HCl (hydrochloric acid)—toxic and corrosive

Procedure

CAUTION! HYDROCHLORIC ACID IS IRRITATING TO THE SKIN AND RESPIRATORY SYSTEM. WEAR SAFETY GLASSES AT ALL TIMES!!

Prepare a flame test wire by obtaining a 10 to15 cm length of Nichrome wire, and bending one end of the wire into a very tiny loop (not more than a millimeter or two in diameter).

Dip the loop end of the flame test wire into a few mL of 6M HCl (to remove any oxides that may be coating the wire). Do not set the wire down any place where it might be contaminated.

Light the Bunsen burner, and open the air-holes of the burner to give a hot flame. Hold the loop end of the flame test wire in the burner flame until it is red hot (this takes only 15 to 20 seconds). Cool the wire in a place where it will not be contaminated.
Obtain 1 or 2 drops of l M NaCl solution on a clean watchglass or in a plastic boat. With the wire loop, pick up a drop of NaCl solution, and place the loop into the burner flame. Observe the color produced and record on the lab report sheet. Caution: The color obtained may only persist for a few seconds before the test solution completely vaporizes.

Re-clean the flame test wire by dipping in 6 M HCl and heating in the burner flame.

Repeat the process for the other available known metal ion solutions: $CaCl_2$, KCl, $BaCl_2$, $SrCl_2$, and $CuCl_2$.

Obtain an unknown sample and record its code number or letter in your lab notebook. Repeat the flame test process and identify which metal ion is contained in the unknown solution.

Note: Since the copper(II) chloride known solution is blue-colored, ALL THE UNKNOWNS HAVE HAD BLUE FOOD COLORING ADDED.

DISPOSAL

Place all the plastic containers and the chemicals into the large beaker labeled Atomic Emissions discarded chemicals.

REMEMBER TO TURN IN THE YELLOW COPY OF YOUR LAB NOTEBOOK TO YOUR TA.

Atomic Emission Spectra
Report Sheet (30 Points)

Name _____   Date _____

Lab Instructor _____   Lab Day _____

Experimental Results

List the flame colors observed for each of the following substances:

NaCl _____

$CaCl_2$ _____

KCl _____

$SrCl_2$ _____

$BaCl_2$ _____

$CuCl_2$ _____

UNKNOWN CODE LETTER _____

IDENTIFICATION OF UNKNOWN _____

Atomic Emission Spectra
Post Lab Questions (40 Points)

Name _____   Date _____

Lab Instructor _____   Lab Day _____

1.   A.   Describe the spectrum you would observe for the emission spectrum of elemental hydrogen gas.

    B.   Describe the spectrum you would observe for the absorption spectrum of elemental hydrogen gas.

2.   List all the possible transitions that can occur between n = 1 and n = 4 in a hydrogen atom. Which ones correspond to lines in the visible region of the spectrum?

3.   WBCN FM broadcasts at a frequency of 104.1 megahertz. What is the wavelength and energy of the radio waves sent out by WBCN. (Note 1 megahertz = 1×10^6 Hz)

4.	If the planet Mars is about 56 million kilometers from Earth, how long would it take for a radio wave sent from a space satellite circling Mars to reach Earth? Assume that radio waves (a form of electromagnetic radiation) travel at the speed of light.

5.	If a laser with a wavelength of 450 nm emits total 7.25×10^{17} photons over a given time period, what is the total energy produced?

6.	A fluorescent mineral absorbs "black light" from a mercury lamp. It then emits visible light with a wavelength 520 nm. The energy not converted to light is converted into heat. If the mineral has absorbed energy with a wavelength of 320 nm, how much energy (in kJ/mole) was converted to heat?

Atomic Emission Spectra
Preliminary Questions (10 Points)

Name _____   Date _____

Lab Instructor _____   Lab Day _____

1. Calculate the wavelength, in nanometers, for the colored line that appears in the spectrum of the hydrogen atom corresponding to an n = 4 to n = 2 electronic transition.

2. Draw a simple energy level diagram for the hydrogen atom, and illustrate with arrows the transitions that give rise to VISIBLE light emissions.

Molecular Weight of a Volatile Liquid

Introduction

The Dumas "vapor density" method, one of the earliest and most simple techniques for determining the FORMULA WEIGHT of a volatile liquid, consists of nothing more than determining the WEIGHT of liquid, which, when VAPORIZED, will fill a given flask at a particular temperature and pressure. A simple variation of the ideal gas law is used to calculate the formula weight: If g represents the mass of liquid required, and M its formula weight, then the number of moles of volatile liquid is just g/M, and the ideal gas law can be written as

$$PV = (g)RT/M$$

You can see from this equation how simple it should be to determine the formula weight, M. All you need to know are the easily measurable PHYSICAL PROPERTIES of the gas (P, V, g, T).

A very simple apparatus is constructed for this experiment (see the sample apparatus), consisting merely of an Erlenmeyer flask covered with a small piece of metal foil, in which a small hole is made with a pin. Several milliliters of the liquid whose formula weight is to be determined are placed in the flask, the foil cover (with the pinhole) is attached, and the flask is placed in a BOILING WATER bath until the liquid has JUST EVAPORATED. At the point where the unknown liquid has just evaporated, the flask is FILLED WITH VAPOR of the liquid. Since the flask is OPEN TO THE ATMOSPHERE via the pinhole, many physical properties of the vapor are known. The PRESSURE of the vapor is just the atmospheric pressure (as read from the barometer). The TEMPERATURE of the vapor is just the temperature of the BOILING WATER. The VOLUME of the flask can be measured by seeing how much water is needed to fill it, and the WEIGHT of the vapor contained can be determined on the balance. The pinhole serves one other function also; any liquid beyond that required to fill the entire flask as a vapor can ESCAPE as the liquid is heated.

Some assumptions are made in this experiment that may affect the results to a small degree. When the flask containing vapor is removed from the boiling water bath, the vapor present will condense as the flask cools back to room temperature; the condensed vapor will occupy a small volume, which had been filled by air before heating. The weight of this small amount of air is negligible, however. Also, the assumption is made that the vapor produced from the liquid will behave as an IDEAL gas: Possibly the worst place to expect a gas to behave ideally is near the temperature where it would liquefy. This factor may introduce a small error into the value calculated for the formula weight; this is taken into account when your results are graded, however.

Summary

The molecular weight of an unknown volatile liquid is determined by the Dumas method, using a flask of known volume, with a small pinhole through a foil cover to permit excess vapor to escape and to equilibrate the vapor to atmospheric pressure.

Supplies

Square piece of aluminum foil; small rubber bands; molecular weight unknown (to which a small amount of iodine has been added as coloring agent)—RECORD THE NUMBER OF THE UNKNOWN ON YOUR REPORT SHEET.

CAUTION! THE HOT WATER BATH USED IN THIS EXPERIMENT MAY SPLASH IF NOT HEATED CAREFULLY. USE BOILING STONES. WEAR SAFETY GLASSES AT ALL TIMES!!

Chemicals

Assume all unknowns are toxic and flammable.

Procedure

Inspect the sample apparatus and assemble your apparatus accordingly.

Fasten the aluminum foil snugly to the top of a clean DRY 125 or 250 mL Erlenmeyer flask with a rubber band. The flask must be COMPLETELY DRY, since any water present will vaporize and affect the results adversely.

With a needle (a piece of metal wire pushed into a cork) make a small pinhole in the center of the foil.

Trim the foil around the edges, so that it will NOT DIP INTO THE WATER bath. (Any droplets of water adhering to the foil will affect the weight of the flask.)

Weigh the apparatus CAREFULLY to the nearest 0.001 g.

Remove the aluminum foil carefully (being sure not to tear it), and add about HALF your unknown liquid (approx. 3 mL). The unknown liquid contains a small amount of iodine, which imparts a color, and makes it easier to note when the liquid has evaporated completely.

Reattach the aluminum foil with the rubber bands.

Heat about 300 mL of water in a 600 mL beaker to a GENTLE boil (use 2 or 3 boiling stones to moderate the boiling).

Immerse the flask containing the unknown in the boiling water as illustrated by the sample apparatus. Have as much of the flask immersed in the boiling water as possible. Use a lightly attached utility clamp on top of the flask to weight it down to ensure that the temperature of the vapor is uniform.

Do not permit the foil cover to come in contact with the water; otherwise, water may be trapped under the foil, giving an incorrect weight.

Heat the flask in the gently boiling water until the liquid has completely evaporated. Then continue heating for approximately 30 seconds longer. Complete evaporation of the liquid can be observed by noticing color changes associated with the iodine.

Determine the temperature of the boiling water bath during this period of heating.

Remove the flask from the water bath, wipe the outside completely DRY and allow it to cool to room temperature (a small amount of liquid will reappear in the flask as it cools, as the vapor produced condenses). Reweigh the flask carefully to the nearest 0.001 g.

Repeat the determination with a portion of the remaining unknown liquid. You do not have to clean out the flask; just add the additional liquid, recover with foil and reheat as above.

DISPOSAL

Pour excess unknown back into your sample vial.

The weight of the flask after it has been heated in the boiling water for this second determination must AGREE with the results of the first determination WITHIN 0.1 g. If the results do NOT agree, use the remainder of your unknown for a third determination.

Finally, using a graduated cylinder as the measuring device, determine the total volume of the Erlenmeyer flask right up to the very top by adding water.

REMEMBER TO TURN IN THE YELLOW COPY OF YOUR LAB NOTEBOOK TO YOR TA.

Molecular Weight of a Volatile Liquid
Report Sheet (30 Points)

Name _____   Lab Day _____

Lab Instructor _____   Date_____

Show calculations at the bottom of this page.

Unknown no. _____

	First Experiment	Second Experiment
Weight of dry apparatus, g	_____	_____
Weight of apparatus plus condensed vapor, g	_____	_____
Weight of condensed vapor, g	_____	_____
Total volume of flask, mL	_____	_____
Total volume of flask, liters	_____	_____
Barometric pressure, mm Hg	_____	_____
Barometric pressure, atm	_____	_____
Temperature of vapor, °C	_____	_____
Temperature of vapor, K	_____	_____
Molecular weight of liquid, g/mole	_____	_____
Average molecular weight of unknown	_____	

Molecular Weight of a Volatile Liquid
Post-Lab Questions (40 Points)

Name _____   Lab Day _____

Lab Instructor _____   Date_____

1.   List two reasons why it is necessary to put a pinhole in the aluminum foil.

2.   Why was it **NOT** necessary to clean and dry the flask before performing your second determination?

3.   A certain liquid has a molecular weight of 100 g/mole. The following data were obtained for a Dumas molecular weight determination:

P = 1.02 atm   V = 0.265 liters   T = 372 K   and
g of condensed vapor = 0.58 g

From this data a molecular weight of 65.5 g/mole was calculated. This value is much less than the expected value of 100 g/mole. What could have accounted for this large error in the molecular weight?

4.   Why was the unknown liquid vaporized in a boiling water bath, rather than at room temperature?

5.   How would the molecular weight for a volatile liquid be affected if the balance you did your weighing on had an error of 2.3 g? (Choose the best answer.)

    A.   The molecular weight would be greater than the correct value.

    B.   The molecular weight would be less than the correct value.

    C.   The molecular weight would be unaffected by this error and it would be the correct value.

6.   It was important that no water gets on the aluminum foil during the heating process. If water does get on the aluminum foil, it must be wiped off before reweighing your flask. A typical drop of water has an approximate volume of 0.05 mL and water has a density of 1.0 g/mL. Calculate the molecular weight of your unknown (using your data from the experiment) if 2 drops of water accidentally remained on the aluminum foil when you reweighed your flask.

Molecular Weight of a Volatile Liquid
Pre-Lab Questions (10 points)

Name _____ Lab Day _____

Lab Instructor _____ Date_____

(Show calculations on the back of the page.)

1.   Define vapor.

2.   If 0.80 g of the vapor of an unknown liquid occupied 280 mL at 100°C and 750 mm Hg, calculate the molecular weight of the liquid.

SOLUBILITY, INTERMOLECULAR FORCES, and POLARITY

A mixture is two or more pure substances physically combined. Some mixtures are uniform throughout, these are called homogeneous or more commonly a solution. Other mixtures are non-uniform throughout and are called heterogeneous. The ability of a substance to mix with a second substance is characteristic of the types of intermolecular forces present in the pure substances.

The ability of a solvent to dissolve a second substance is in most cases not limitless. The maximum amount of solute that can be dissolved in a given amount of solvent is called the solubility. For example, 35.7 g of sodium chloride can be dissolved in 100 g of water at 25°C. If we attempt to dissolve more than this quantity of sodium chloride in the water, the excess solid would drop to the bottom of the beaker and remain undissolved. A solution in which the maximum quantity of solute is dissolved is called a saturated solution. The terms solubility and a saturated solution describe the same concept. The difference is that the solubility is a quantitative measurement, while saturated solution is a qualitative one. In a saturated solution, we do not care how much solid has been dissolved, but that there remains undissolved solid that cannot be dissolved at that temperature no matter how much shaking or stirring is done.

Liquids also behave similarly when dissolved in other liquids. Ethyl alcohol and water mix in all proportions. Two liquids that mix in all proportions are said to be miscible. Other liquids do not mix at all in any appreciable quantities, such as oil and water. These liquids are said to be immiscible. Solutions that only partially mix or do not mix at all will separate into two layers, with the less dense layer on top.

Temperature also can affect the degree of solubility. For an endothermic process (that is, the dissolving process absorbs energy from the surroundings, making the surroundings colder) the solubility increases with increasing temperature. The opposite is true for an exothermic process (one that gives off energy to the surroundings, making the surroundings warmer); that is, an increase in temperature decreases the solubility. This concept will not be examined in this experiment.

One general rule that can be used to determine whether two substances will be miscible is the statement "like dissolves like". This means that substances with similar intermolecular forces should form a solution. For example, two polar substances should form a solution (like water and ethyl alcohol) or two nonpolar substances should form a solution (such as benzene and carbon tetrachloride). We should expect that a polar substance like water and a nonpolar substance like oil will not form a solution. In fact we know that oil floats on top of water.

At this point, you might ask the question—why does sodium chloride dissolve in water? The bonding in sodium chloride (NaCl) is ionic. That is, an electron has been transferred from the sodium atom to the chlorine atom. When this happens, the sodium atom acquires a positive charge (Na^+, because it has one more proton than electron) and the chlorine atom becomes negatively charged (Cl⁻, because it has one more electron than proton). The positive and negative charges hold the substance together in a three-dimensional arrangement of positive and negative charges called a lattice.

When the solid is put in water, the positive ends of the water molecule dipoles are attracted to the chloride ions (Cl⁻) and the negative ends of the water dipole are attracted to the positive sodium ions. Since more than one water molecule can act on an individual ion, the water molecules can collectively pull the positive and negative ions out of their lattice positions, eventually surrounding the ion. At this point the ion is now considered to be dissolved. Eventually the solution reaches a state where the maximum solubility is reached. Some ionic substances are basically insoluble in water. In these substances the forces holding the ions together are too strong and the water molecules are unable to pull the ions out of their lattice positions. The solid remains intact and the substance is insoluble.

We can generalize this information by looking at interactions between various substances.

Solute-solute attractions—attractive forces between two solute particles or molecules

Solvent-solvent attractions—attractive forces between two solvent molecules

Solute-solvent attractions—attractive forces between a solute particle or molecule and one or more solvent molecules

If solute-solvent attractions are greater than or roughly the same as the solute-solute or solvent-solvent attractions, we should expect solution formation to occur in measurable amounts.

If solute-solute attractions or solvent-solvent attractions are much greater than solute-solvent attractions, we should expect the solubility to be extremely small or zero.

The ability of a solute to be dissolved in a solvent can easily be related to the bonding arrangement of the atoms in both the solute and solvent. Hence the expression "like dissolves like" refers to the fact the polar solutes are generally soluble in polar solvents and nonpolar solutes are generally soluble in nonpolar solvents. (An ionic bond can be considered to be so polar that there are full charges and not just partial charges present.)

In this experiment you will work in groups of three students. Each person will be responsible for determining the solubility of a group of compounds (solutes) in various solvents as follows:

Group Member 1:

    Solutes: Methanol, isopropanol (isopropyl alcohol), n-butanol (n-butyl alcohol), cyclohexanol, ethylene glycol, and glycerol

    Solvents: Distilled water, ethanol (ethyl alcohol), hexane and acetone

Group Member 2:

    Solutes: Sodium chloride, copper(II) sulfate, sodium acetate, potassium bromide, sucrose, and iodine

    Solvents: Distilled water, ethanol (ethyl alcohol), hexane and acetone

Group Member 3:

    Solutes: Naphthalene, pentane, vegetable oil, acetic acid, benzoic acid, and stearic acid

    Solvents: Distilled water, ethanol (ethyl alcohol), hexane and acetone

DISPOSAL: Do not dispose of any of your solutions by pouring them down the drain. Put your waste into the proper disposal bottle as follows:

Group Member 1:   Pour all solutions into the waste alcohol container.

Group Member 2:   Pour all solutions containing sodium chloride, copper(II) sulfate, potassium bromide, and sodium acetate into the waste salts container.
Pour all solutions containing iodine into the waste iodine container.
Pour all solutions containing sucrose into the waste sugar container.

Group Member 3:   Pour all solutions containing acetic acid, stearic acid, and benzoic acid into the waste carboxylic acid container.
Pour all solutions containing naphthalene and pentane into the waste hydrocarbon container.
Pour all solutions containing vegetable oil into the waste vegetable oil container.

PROCEDURE

Note: It may be difficult to differentiate between partially soluble and insoluble. The best procedure is to add only a few drops at a time and then agitate the test tube looking for two layers.
When adding a solid, use only a very small quantity. Since solvent quantity is also very small, it takes only a small amount of solute to reach the solubility limit.

A.   Liquids mixed with other liquids

    To **six 10 x 75** test tubes add about 1 mL (about 20 drops) of distilled water to each one. To each test tube add drop by drop the solutes listed in your particular part of this experiment up to a maximum of 20 drops (about 1 mL) or until the mixture turns cloudy. This cloudiness indicates that the two liquids are no longer soluble and one should see two distinct layers upon sitting. The number of drops required for layer formation to occur indicates the solubility limit of the substance in the water.

If any of your solutes are solids, follow the procedure given below.

B.   Solids mixed with a liquid

To **a 10 x 75 mm** test tube, add 1 mL (about 20 drops) of distilled water and a small amount of the solid to the test tube. **Never handle chemicals by hand: Use a spatula or spoon.** Tap the side of your test tube near the bottom to mix the solution.

When you have completed testing the solubility of your six solutes in distilled water, repeat the procedure with the other three solvents.

Dispose of your waste solutions in the proper containers.

When you have completed the testing of each solvent, fill in the table in your lab notebook for your group of solutes with an **s** for **completely soluble**, a **p** for **partially soluble,** and an **i** for **insoluble**. Transfer the data of your partner's data also into your lab notebook.

BE SURE TO PICK UP THE TWO INFORMATION SHEETS.

The first one gives you information about the bonding in each compound used in the experiment.

The second one tells you the melting point, the boiling point, and the molar mass of each substance used in the experiment.

Observations
small solvent + polar ⟶ dissolves polar + sma
                           not large + polar

Slight release of gas when solute dissolves

∴

124

Cryoscopic Determination of Molecular Weight

Introduction

COLLIGATIVE properties of solutions are those properties that depend only on the CONCENTRATION of particles in the solution, not on the nature of the particles. For example, a 1.0 molal solution of the sugar glucose in water should have the SAME colligative properties as a 1.0 molal solution of the sugar sucrose (or any other nonionizable solute). Colligative properties include such things as boiling point elevation, vapor pressure lowering, osmotic pressure, and, as in this experiment, freezing point lowering.

Whenever a solute is added to a solvent, the TEMPERATURE at which the SOLUTION freezes is LOWERED, relative to that of the SOLVENT ALONE. For example, considering 1.0 kilogram of water as the solvent, the freezing point is lowered by 1.86°C for EACH MOLE of solute added. When the concentration of a solution is expressed in terms of the MOLES OF SOLUTE PER KILOGRAM OF SOLVENT (i.e., the molality, m), the CHANGE in freezing point, ΔT, relative to the freezing point of the pure solvent, is given by

$$\Delta T_f = K_f m$$

where K_f is a proportionality constant (called the "molal freezing point depression constant" for the solvent under consideration).

In this experiment, you will prepare a SOLUTION of an unknown material in the solvent NAPHTHALENE very carefully by WEIGHT, and will then measure the CHANGE in freezing point of this solution (relative to the freezing point of pure naphthalene). From the change in freezing point, and from the constant, K_f, for the naphthalene solvent, you will be able to calculate the actual MOLALITY of your solution. By knowing the makeup of the solution by weight, you will be able to calculate the molecular weight of the unknown material. The constant K_f for naphthalene is 6.9°C/molal.

The freezing point of a PURE SUBSTANCE is accurately determined by recording temperature/time data for the cooling of the molten substance; i.e., the substance is melted completely, and then its temperature is taken at regular time intervals as the substance cools, until it solidifies completely. For a typical substance, the temperature will drop regularly with time as the substance is cooled. Then, as the substance begins to freeze, the temperature will remain CONSTANT for several minutes. Finally, the temperature will begin to drop again with time once complete solidification has taken place. THE TEMPERATURE THAT REMAINS CONSTANT FOR SEVERAL MINUTES IS THE FREEZING POINT OF THE SUBSTANCE. This sort of temperature/time data is best treated GRAPHICALLY (see Figure 1), where the FLAT portion of the curve represents the freezing point.

The freezing point of a SOLUTION can be determined by the same sort of temperature/time technique, although the temperature will probably not remain entirely constant as the solution freezes (as the solution freezes, the relative ratio of solute to solvent changes, thereby changing the molality, and hence the freezing point). The graphical treatment of data for the solution (see Figure 2) does work, however, since the freezing point of the solution is indicated by an abrupt CHANGE IN THE SLOPE of the curve as freezing begins (and again when freezing ends).

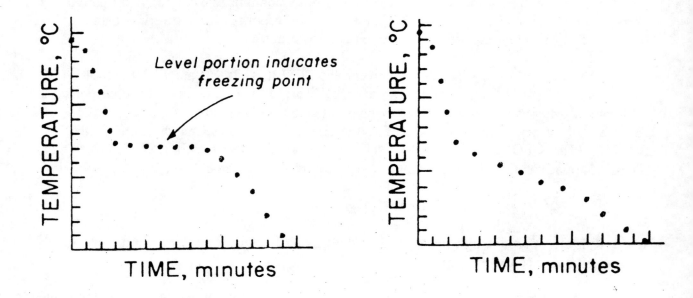

Figure 1: Cooling Curve for a Pure Solvent   **Figure 2: Cooling Curve for a Solution**

WHEN YOU HAVE FINISHED DETERMINING THE FREEZING POINTS OF THE UNKNOWN SOLUTIONS IN THIS EXPERIMENT, DISCARD THE SOLUTIONS IN THE SPECIAL CONTAINERS PROVIDED—NOT IN THE SINKS!!

Summary

The molecular weight of an unknown solid organic compound is determined by the freezing point depression (cryoscopic) method, using naphthalene as solvent. The molal freezing point depression constant, K_f, for naphthalene is 6.9°C/molal. This experiment may be done by students working in pairs—one person reading temperatures, and the other recording temperature/time data.

Supplies

Pure naphthalene; molecular weight unknown; watch with sweep second hand.

RECORD THE NUMBER OF THE UNKNOWN IN YOUR LAB NOTEBOOK.

Procedure

CAUTION! USE CARE IN INSERTING YOUR THERMOMETER THROUGH THE RUBBER STOPPER. ACETONE IS HIGHLY FLAMMABLE—NO FLAMES PERMITTED DURING ITS USE!

WEAR SAFETY GLASSES AT ALL TIMES!

(1) Freezing Point of Pure Naphthalene

Using glycerin as lubricant (if your rubber stopper is not already split) insert the TOP of your thermometer (CAUTION!!) through a rubber stopper in such a way that it can be read from 65 to 90°C.

Weigh approximately 5 to 6 g (record the exact weight) of PURE NAPHTHALENE into a clean, dry test tube (your small one 8 x 150 mm).

MELT the naphthalene COMPLETELY by clamping the test tube VERTICALLY in a boiling water bath in a 600 mL beaker.

When the naphthalene has melted COMPLETELY and the temperature of the naphthalene has reached 95°C, remove the beaker of boiling water, and suspend the thermometer from above to dip halfway into the melted naphthalene WITHOUT TOUCHING THE SIDES OR BOTTOM OF THE TEST TUBE. See Figure 3 (next page). Use the rubber hot hands or several pieces of paper towel to remove the beaker.

When the naphthalene temperature has dropped to 90°C, start taking temperature readings (to the nearest 0.2 degree—estimate between degree marks) EVERY 30 SECONDS until the temperature has dropped to below 65°C (at which temperature the naphthalene should have frozen completely).

Plot your temperature/time data on a piece of graph paper, taking the zero of time to correspond to when the temperature was 90°C. The HORIZONTAL flat portion of the curve defines the freezing point of the naphthalene.

If your graph gives poor results, remelt the naphthalene to above 90°C and repeat the determination.

133

If your graph is reasonable, continue on to part 2.

(2) Freezing Point of Solution of Unknown in Naphthalene

Suspend your thermometer back in a beaker of water. Heat the beaker and contents to remelt the naphthalene. Remove the thermometer and stirrer from the test tube. Allow any liquid naphthalene to drip back into your test tube. Wait for the naphthalene to solidify on the thermometer and the stirrer, then carefully clamp your thermometer to the ring stand and gently place your stirrer on a piece of paper towel on your benchtop. Make sure that none of the solid becomes dislodged. You will use all of the naphthalene for this part of your experiment.

Remove your test tube from the hot water bath and place it into a cold water bath to speed up the solidification process. Place the test tube in the bath so it is upright and make sure that the beaker cannot tip over. When the test tube is at or near room temperature, remove it from the cold water and completely dry it before you weigh it again.

Weigh the dried test tube from part 1 to the nearest 0.01 g.

Obtain an unknown from the supply table in your lab. Record the unknown code in your lab notebook.

Weigh out about 1.0 g (exact weight to the nearest 0.01 g) of your unknown onto a piece of weighing paper and then transfer it to your test tube and reweigh it. Determine the mass of unknown in your sample.

Reassemble the apparatus and repeat the procedure for determining the melting.

Melt the mixture in a boiling water bath and STIR THE MOLTEN MIXTURE with the thermometer and the wire stirrer (if available) THOROUGHLY to ensure that the mixture is completely homogeneous.

When the temperature of the molten mixture has reached 95°C, remove the beaker of boiling water, adjust the thermometer so that it hangs in the MIDDLE of the solution, and determine the cooling curve as done for naphthalene alone.

Usually the freezing point of the solution is the first point of deflection of the cooling curve (see Figure 2). However, SUPERCOOLING can occur (i.e., the mixture DROPS MOMENTARILY BELOW ITS ORDINARY FREEZING POINT until nucleation of solid formation can occur).

If this happens to you, your lab instructor will show you how to EXTRAPOLATE your data back to the temperature axis to get the freezing point. If your data give a really poor graph, remelt the sample and do a second determination.

Reassemble your apparatus so that your test tube is in the beaker of water. Heat the water to melt the mixture. When the solution has melted, again hold the stirrer and thermometer above the test tube and allow any liquid to drip back into the test tube. When no more liquid drips off, you can put the stirrer and thermometer on your bench. Next unscrew the clamp from the ring stand, and quickly take the clamp with the attached test tube over to the hood and pour the contents into the waste container. Bring your test tube to the stockroom and leave your test tube with us and we will give you a clean replacement.

Calculate the molecular weight of your unknown, and attach the cooling curve graphs you have drawn to your report.

Remember to turn in the carbon copy from your lab notebook to your TA.

CALCULATIONS

1.   Plot time/temperature data for naphthalene.

      Determine the freezing point for pure naphthalene.

2.   Plot time/temperature data for your naphthalene-unknown mixture.

      Determine the freezing point of your solution.

      Using the K_f for naphthalene (6.9°C/m) and your graph, calculate the following quantities.

      i)   The freezing point depression of the solution.
      ii)   The molality of your unknown.
      iii) The molar mass of your unknown.

3.   Remember to turn in both graphs with your lab report.

Inorganic Synthesis—Preparation of Sodium Thiosulfate Pentahydrate

Introduction

In this experiment, you will prepare an important substance which finds many uses, both in the Chemistry laboratory, and in everyday life: sodium thiosulfate pentahydrate, $Na_2S_2O_3 \bullet 5H_2O$.

SODIUM THIOSULFATE PENTAHYDRATE is prepared by dissolving powdered elemental SULFUR in a solution of SODIUM SULFITE. Since powdered sulfur is very "dry", a small amount of DETERGENT is added to assist it in dissolving:

$$S + Na_2SO_3 + 5H_2O \rightarrow Na_2S_2O_3 \bullet 5H_2O$$

In the laboratory, sodium thiosulfate is used in the analysis of IODINE. The thiosulfate ion reacts QUANTITATIVELY (i.e., completely, quickly, and stoichiometrically) with elemental iodine in aqueous solution. Thus, STANDARD solutions of thiosulfate ion can be used to TITRATE unknown samples containing iodine (or iodide ion, which can be oxidized to iodine before titration). The disappearance of the iodine's BROWN COLOR can be used as an indication of the endpoint of the titration (also, STARCH is commonly added to iodine samples to be titrated to intensify the iodine's color). However, this experiment does not involve any titrations.

$$2\,S_2O_3^{2-} + I_2 \rightarrow S_4O_6^{2-} + 2\,I^-$$
$$\text{brown} \qquad\qquad \text{colorless}$$

Sodium thiosulfate is commonly used in PHOTOGRAPHY (where it is known as photographers' "hypo"). Common photographic film is coated with a SILVER SALT (most often silver bromide, AgBr). When the silver salt is exposed to light through the camera's lens, some of the silver ion is reduced to METALLIC SILVER by the light (the extent to which this happens is a function of the INTENSITY OF THE LIGHT striking the film emulsion). When the film is DEVELOPED, any silver bromide still present in the emulsion must be REMOVED (or it will darken when exposed to light subsequently). The thiosulfate ion forms WATER SOLUBLE complex ions with SILVER ION (but NOT with METALLIC silver), and can be used to REMOVE the ordinarily insoluble AgBr from the film. The metallic silver left behind then forms the negative film image.

$$AgBr + 2S_2O_3^{2-} \rightarrow Ag(S_2O_3)_2^{3-} + Br^-$$

Since sodium thiosulfate is the salt of the weak acid $H_2S_2O_3$ (making sodium thiosulfate a base), the salt will react with strong acids such as HCl. The product, $H_2S_2O_3$, is unstable, and disproportionates to elemental sulfur, sulfur dioxide, and water:

$$S_2O_3^{2-} + 2H^+ \rightarrow H_2S_2O_3 \rightarrow H_2O + SO_2 + S$$

When prepared from aqueous solution, sodium thiosulfate ordinarily crystallizes out as the PENTAHYDRATE. This substance, however, loses the waters of crystallization on standing. This property of hydrates is called "efflorescence," and is quite common for salts crystallized from aqueous solution. On a very dry day, when there is little moisture in the atmosphere, sodium thiosulfate pentahydrate readily loses its waters of crystallization, and a noticeable change takes place in the salt's appearance and texture.

Summary

Sodium thiosulfate pentahydrate is prepared by heating sulfur with a solution of sodium sulfite in the presence of a small amount of detergent (which promotes the dissolving of the sulfur). Reactions illustrating the laboratory and commercial uses of thiosulfate are performed on the product.

Supplies

Sodium sulfite, sulfur; diluted detergent solution; ice; 0.1 M iodine solution; 0.1 M NaCl; 0.1 M silver nitrate; 6 M HCl; suction flask; Buchner funnel and filter paper to fit.

CHEMICALS

Sodium sulfite (Na_2SO_3)—skin irritant
Sulfur—skin irritant
Silver nitrate—will stain skin (but does no harm)
Iodine (I_2)—toxic, eye irritant, and somewhat corrosive to skin
Potassium iodide (KI)—no major health risks
Hydrochloric acid (HCl)—toxic, corrosive, and will burn skin
Sodium chloride (NaCl)—no major health risks
Sodium thiosulfate ($Na_2S_2O_3 \bullet 5H_2O$)—no major health hazards

Procedure

CAUTION! THE PRODUCT SOLUTION MAY BEGIN TO SPLATTER IF HEATED TOO STRONGLY. STOP HEATING IF SPLATTERING OCCURS. WEAR SAFETY GLASSES!!

Weigh 12.6 g of sodium sulfite and 3.5 g of powdered sulfur into clean, dry beakers. NOTE: Do not weigh chemicals directly on the balance pans. Use weighing paper or filter paper or a beaker.

Add these reactants, plus around 1 mL (e.g., one dropperful) of diluted detergent solution, to 60 mL of water in a 150 mL beaker.

Add two boiling stones, and heat the mixture to a GENTLE boil, STIRRING to dissolve the sulfite and to promote wetting of the powdered sulfur.

Continue heating for 15 to 20 minutes until the solution when tested with pH paper from the laboratory, is neutral or only slightly basic. (The pH paper should test light to moderately dark green.) The mixture will have been fairly strongly basic (pH paper will be dark blue) when you began heating, and it becomes less basic during the heating period.

Filter the hot solution through a GLASS GRAVITY FUNNEL into a clean 150 mL beaker (to remove unreacted material).

The filter paper with the unreacted sulfur should be discarded into the waste sulfur beaker.

Add two boiling stones, and EVAPORATE water from the solution by GENTLE boiling (stirring occasionally with the thermometer). Heat until the volume of the solution is reduced to 20 to 25 mL (at this point, the concentration of dissolved material in the solution is such that the boiling point should be between 106 and 108°C). DO NOT EVAPORATE THE SOLUTION SO FAR THAT SPLATTERING BECOMES A PROBLEM.

Remove the burner flame and allow the beaker to cool somewhat in the air.

Then cool the beaker, first in cold water, then in ice, until the temperature reaches 5 to 7°C.

If crystals have not formed at this point, the solution is probably supersaturated. In this case, nucleate crystallization by rubbing the bottom of the beaker with a glass stirring rod.

When crystallization is complete, filter the crystals by suction, using a BUCHNER funnel. Transfer the crystals to an 11 cm piece of filter paper and allow them to dry.

NOTE: If all of your liquid seems to disappear and the mixture starts to become very thick, filter it immediately. If you do not, the mixture will form a solid, which will be very difficult to remove from your beaker.

While the crystals are drying, perform the qualitative tests outlined on the report sheet, using small portions of your product.

The liquid remaining in the filter flask (the filtrate) can be poured down the drain.

QUALITATIVE TESTS

a.   Add 2 drops of 0.1 M NaCl and 2 drops of 0.1 M silver nitrate to about 5 mL of water in a test tube to produce AgCl. Add a few crystals of your reaction product and shake. Describe your observations in your lab notebook.

b.   Add 1-2 drops of 0.1 M iodine solution to about 5 mL of water in a test tube. Add a few crystals of your product and describe your observations in your lab notebook.

c.   Dissolve a small amount of your product in about 5 mL of water in a test tube. Add about 1 mL of 6 M HCl. Warm the solution and sniff it cautiously. Describe your observations in your lab notebook.

DISCARD ALL WASTE INTO THE IONORGANIC SYNTHESIS WASTE BOTTLE.

Finally, weigh your dried product and indicate the actual and percentage yields.

Discard the sodium thiosulfate pentahydrate product and the filter paper in the beaker labeled WASTE SODIUM THIOSULFATE.

REMEMBER TO TURN IN THE YELLOW COPY OF YOUR LAB NOTEBOOK TO YOUR TA.

$NaCl + AgNO_3 \longrightarrow$

Kinetics of an Iodine Clock Reaction

Introduction

Chemical kinetics is the study of the RATES at which chemical reactions proceed. For a general reaction

$$A + B \rightarrow \text{Products}$$

it is found experimentally that the rate is PROPORTIONAL TO THE CONCENTRATIONS of the reactants A and B

$$\text{Rate} = k[A]^x[B]^y$$

This expression is called the RATE LAW for the reaction. The exponents x AND y are called the ORDERS of the reaction with respect to the concentrations of A and B, respectively (and reflect, in many cases, the actual number of A and B molecules that must physically interact during the crucial, or rate-determining, step of the process (see Note l) and k is called the SPECIFIC RATE CONSTANT for the reaction. (It provides the proportionality needed between the observed rate and the concentrations of reactants.)

The ORDER with respect to a given reactant in a process CAN ONLY BE DETERMINED EXPERIMENTALLY. A simple technique can be used to accomplish this. To determine the order of reaction with respect to a particular reactant, several experiments are done using DIFFERENT concentrations of the REACTANT IN QUESTION; however, the concentrations of all OTHER SUBSTANCES are kept CONSTANT during these experiments. Thus, any variation in the observed rate will be due to the change in concentration of the reactant under study. This method is called THE INITIAL RATE METHOD.

For example, if the concentration of A in the above reaction is DOUBLED, relative to an initial experiment, and the observed rate of reaction also DOUBLES (with B concentration kept the SAME in both experiments), then the exponent x of [A] in the rate law above MUST EQUAL 1 (since $2^1 = 2$). If, for example, in a subsequent set of experiments with [A] kept constant, it is observed that the rate of reaction QUADRUPLES when [B] is DOUBLED, then the exponent y of [B] must equal 2 (since $2^2 = 4$). Simple analysis of such rate/concentration data enables one to determine orders of reaction quite simply in many cases. Ordinarily, the order is confirmed graphically by a method discussed below.

In this exercise, we want to CONFIRM that the dependence of the overall rate of reaction for the following process is FIRST order in iodide ion concentration [I^-].

$$S_2O_8^{2-} + 2I^- \rightarrow 2SO_4^{2-} + I_2$$

persulfate   iodide   sulfate   iodine

That is, we want to show that the exponent y equals 1 in the rate law, which can be written for the above reaction.

We will measure the TIME REQUIRED for this reaction to occur at different concentrations of iodide ion, keeping persulfate ion concentration constant throughout, since the time required for reaction is, of course, related to the rate of the reaction. To simplify the experiment (in addition to keeping the initial concentration of persulfate ion constant in all runs), the extent of reaction is also kept small and constant by adding a small amount of sodium thiosulfate to each reaction mixture (see Note 2). A small amount of starch is also added to each reaction mixture to indicate production of iodine (iodine produces an intense blue-black color with starch, which is much easier to see). Increasing amounts of non-reactive KCl are added to the several reaction mixtures to compensate for less KI being present (the total number of ions present should be kept constant during the several reaction runs for valid comparisons of relative rates). This is always necessary in reactions involving ionic species.

We would like to GRAPHICALLY confirm the first order dependence of the overall rate of the reaction on iodide ion concentration. To do this, the rate law

$$\text{Rate} = k[S_2O_8^{2-}]^x[I^-]^1$$

can be written more simply as

$$\text{Rate} = k'[I^-]^1$$

since $[S_2O_8^{2-}]$ is kept constant throughout, and can be included in the new constant k'. But the rate of a reaction represents the CHANGE in concentration, conc., of one of the reacting species over a time interval, t. That is,

$$\text{Rate} = \frac{\Delta \text{Conc}}{\Delta \text{time}} = k'[I^-]^1$$

Since the extent of reaction, conc., is kept small and constant in this experiment (see Note 2), then the quantity $1/\Delta$time is proportional to $[I^-]^1$

So a graph of $1/t$ versus $[I^-]$ should be a STRAIGHT LINE. Data will be collected to see if this is the case.

Qualitatively, we expect an increase in temperature to speed up a reaction. This effect will be observed also.

Note 1:   The number of molecules that must actually come together physically during the rate-determining step of a reaction is more aptly called the molecularity of the reaction.

Note 2:   The thiosulfate ion added reacts instantaneously with the iodine as t is formed. The net ionic equation for this is

$$I_2 + 2S_2O_3^{2-} \rightarrow 2I^- + S_4O_6^{2-}$$

When the added thiosulfate is used up in the above reaction, iodine formed in the reaction under study immediately imparts a blue-black color to the mixture (because of the starch present). Since the extent of reaction at the starch endpoint is very small, concentrations of iodide and persulfate ions are essentially the same as the initial concentrations. This approach to kinetic experimentation is called the "method of initial rates."

Supplies

1.0 M KI; 1.0 M KCl; 0.10 M $(NH_4)_2S_2O_8$; 0.030 $Na_2S_2O_3$ (with added starch).

Chemicals

Sodium thiosulfate ($Na_2S_2O_3$)—no major health hazards
Potassium iodide (KI)—no major health hazards
Potassium chloride (KCl)—no major health hazards
Ammonium persulfate (($NH_4)_2S_2O_8$)—strong oxidizer
Starch—no major health hazards

Procedure

(1) General

You should work in pairs on this experiment. A watch with a sweep second hand will be helpful. (Stop watches are available from the stockroom.) In preparing solutions, take particular care that the iodide solution and the persulfate solution are not contaminated, one with the other. Measure all volumes carefully and label or mark all containers.

(2) Preparation of Solutions

POTASSIUM IODIDE: Obtain 10 mL of 1.0 M KI and dilute it to 100 mL with deionized water (that is, add 90 mL of deionized water). Store the resulting 0.10 M KI solution in a clean 250 mL Erlenmeyer flask.

POTASSIUM CHLORIDE: Dilute 5 mL of 1.0 M KCl to 50 mL with deionized water. Store the resulting 0.10 M KCl solution in a clean 125 mL Erlenmeyer flask.

SODIUM THIOSULFATE: Obtain 10 mL of 0.030 M sodium thiosulfate solution (which contains a small amount of starch indicator) and dilute it with deionized water to 100 mL. Store the resulting 0.003 M solution in a clean 400 mL beaker.

AMMONIUM PERSULFATE: The ammonium persulfate solution provided by the stockroom is already at the concentration needed for this experiment. Obtain 125 mL of the available 0.10 M ammonium persulfate solution and store in clean 400 mL beaker.

(3) **Kinetic Runs**

First, determine the temperature of all solutions, rinsing and drying the thermometer between temperature measurements (to avoid contamination of the solutions). Temperatures should be within 1°C of each other; if they are not, warm or cool the containers as necessary.

Use three labeled graduated cylinders for measuring solutions: one for KI and KCl; one for thiosulfate; and one for persulfate. The order of mixing is important: Measure into a freshly cleaned (and dried) 250 mL beaker, the iodide solution, the KCl solution (if any), and the thiosulfate solution as specified as in **Table I** (shown on the next page). Measure the proper amount of persulfate solution into its graduated cylinder.

Making note of the time, add the persulfate solution all at once to the solutions in the beaker. STIR THE SOLUTION for 10 to 15 seconds immediately to ensure homogeneity, then OBSERVE THE REACTION MIXTURE CAREFULLY for the appearance of color. Record the time required for the color change to the NEAREST SECOND. Stop timing at the first sight of the blue color forming. Start your timing as soon as you add the persulfate solution (not after mixing).

Carry out all four kinetic runs as outlined in Table I (below), recording reaction time data in the table in your lab notebook. Calculate the molar concentration of I^- in the reaction mixture for each run. Using a piece of graph paper, construct two graphs based on your data.

In the first graph, plot $[I^-]$ vs. reaction time, t.

In the second graph, plot $[I^-]$ vs. 1/t. A straight line in the second graph indicates, as discussed earlier, that the reaction is first order with respect to the iodide concentration. If a data point for one of your runs appears to be in error (from inspection of the second graph), you may wish to repeat that run. Attach both graphs to your report.

Run Number	0.1 M KI	0.1 M KCl	0.003 M Thiosulfate	0.10 M Persulfate
1	20 mL	0 mL	10 mL	20 mL
2	15 mL	5 mL	10 mL	20 mL
3	10 mL	10 mL	10 mL	20 mL
4	5 mL	15 mL	10 mL	20 mL

(4) Effect of Temperature

Select a reaction mixture corresponding to one of the previous runs. Heat (or cool) all reaction solutions to a temperature approximately 10°C warmer (or cooler) than that of the initial set of runs. Carry out the run as before and determine the reaction time.

Dispose of all solutions into the bottle labeled WASTE IODINE CLOCK SOLUTIONS.

REMEMBER TO TURN IN THE YELLOW COPY OF YOUR LAB NOTEBOOK TO YOUR TA.

Determination of an Equilibrium Constant

Introduction

Many chemical reactions, especially organic ones, are REVERSIBLE. If the reaction is carried out in a CLOSED SYSTEM, eventually an EQUILIBRIUM will be established among all the participating substances. For a general reaction

$$aA + bB \rightleftharpoons cC + dD$$

The CONCENTRATIONS of the four participating substances are related AT EQUILIBRIUM by the EQUILIBRIUM CONSTANT, K_c

$$\frac{[C]^c[D]^d}{[A]^a[B]^b}$$

which is a RATIO of PRODUCT concentrations (raised to POWERS given by the COEFFICIENTS of the chemical equation for the reaction) over REACTANT concentrations (also raised to powers) given by the coefficients of the balanced chemical equation.

Once the VALUE of the equilibrium constant has been determined for a given reaction at a particular temperature, predictions can be made about effects of concentration changes on the reaction, since the NUMERICAL VALUE of K_c does NOT CHANGE unless the temperature is changed. Examples of this sort of calculation are to be found in your textbook.

In this experiment, we will be determining the value of the equilibrium constant for the ESTERIFICATION reaction that takes place between ISOPROPYL ALCOHOL (2-propanol) and ACETIC ACID (ethanoic acid):

$$C_3H_7OH + CH_3COOH \rightleftharpoons CH_3COOC_3H_7 + H_2O$$

isopropyl   acetic   isopropyl   water
alcohol   acid   acetate

The reaction is CATALYZED by sulfuric acid: The sulfuric acid allows the equilibrium to be reached more QUICKLY, but the sulfuric acid is NOT CONSUMED in the reaction (the amount of sulfuric acid will be the SAME at the START of the reaction, and once EQUILIBRIUM is reached). The sulfuric acid will NOT CHANGE the relative amounts of products produced, or reactants remaining, once the equilibrium is reached.

You might think it necessary to determine experimentally the concentration of EACH of the four substances involved in the expression for K. This is NOT necessary in this experiment. If we start the reaction with EQUAL AMOUNTS (on a mole basis) of isopropyl alcohol and acetic acid, and then just measure how much of ONE of the four substances is present after reaction has occurred, we can CALCULATE WHAT THE EQUILIBRIUM CONCENTRATIONS OF THE OTHER SUBSTANCES MUST BE from the stoichiometry of the chemical reaction.

For example, suppose we start with a mixture that is 10.0 M in isopropyl alcohol and also 10.0 M in acetic acid, and find that, at equilibrium, the mixture is now only 3.0 M in acetic acid. Therefore, the mixture must also be 3.0 M in isopropyl alcohol (since we started with equal amounts of isopropyl alcohol and acetic acid, and the reaction occurring is of 1:1 stoichiometry). The CHANGE IN CONCENTRATION OF REACTANTS is reflected in the CONCENTRATIONS OF PRODUCTS that build up: Isopropyl acetate and water should now be 7.0 M each. The value of the constant would thus be:

		ICE	TABLE	
	Acetic Acid	Isopropyl Alcohol	Isopropyl Acetate	Water
Initial Concentrations	10.0	10.0	0	0
Change in Concentration	- X	- X	+ X	+ X
Equilibrium Concentration	10.0 - X = 3.0	10.0 - X = 3.0	X = 7.0	X = 7.0

$$\frac{[7][7]}{[3][3]} = 5.56$$

In this experiment, we determine the amount of ACETIC ACID (in moles/liter) present in the INITIAL mixture, and again in the mixture AT EQUILIBRIUM. This is done by TITRATING the acetic acid with standard 0.200 M NaOH to a phenolphthalein endpoint. The acidic hydrogen of the acetic acid molecule REACTS with the hydroxide ion of NaOH in a 1:1 stoichiometric ratio

$$H^+ + OH^- \rightarrow H_2O$$

166

For example, suppose a 1.00 mL sample of the initial acetic acid/isopropyl alcohol mixture requires 35.00 mL of 0.200 M NaOH to reach the endpoint of a titration. The moles of NaOH used in the titration is

$$(0.03500 \text{ liters})(0.200 \text{ moles/liter}) = 0.00700 \text{ moles NaOH}$$

The moles of acetic acid present in the sample taken must equal this quantity.

$$\text{moles acetic acid} = \text{moles NaOH used} = 0.00700 \text{ moles}$$

Since this number of moles was contained in 1.00 mL of the original reaction mixture, the CONCENTRATION of acetic acid (in moles/liter) in the original mixture must be

$$0.00700 \text{ moles}/0.00100 \text{ liters} = 7.00 \text{ M}$$

There is one complication, however. The sulfuric acid added to the initial reaction mixture as a catalyst WILL ALSO REACT WITH NaOH when a sample of equilibrium mixture is titrated to determine acetic acid concentration. A BLANK is prepared containing the SAME AMOUNT OF SULFURIC ACID per milliliter as the reaction mixture, and a titration is performed to determine how many milliliters of standard 0.200 M NaOH are needed to react with the sulfuric acid present.

Then, when the equilibrium reaction mixture (containing sulfuric acid) is titrated, the number of milliliters needed to react with the sulfuric acid present can be SUBTRACTED from the total volume of NaOH used to get the amount of NaOH used to react with JUST THE EQUILIBRIUM ACETIC ACID. From this CORRECTED volume of NaOH used to titrate the equilibrium acetic acid, the concentration of acetic acid (in moles/liter) in the mixture AT EQUILIBRIUM can be calculated as above.

Supplies

Fifteen grams (14.2 mL, 0.25 moles) glacial acetic acid (CAUTION); 15 g (19.2 mL, 0.25 moles) anhydrous isopropyl alcohol; standard 0.200 M NaOH solution; phenolphthalein indicator solution; dropper bottle of concentrated sulfuric acid (CAUTION: HOLD BOTTLE WITH TOWEL); 1 mL TD pipet; buret and clamp; pipet bulb.

Chemicals

Isopropyl alcohol—flammable liquid, toxic by ingestion and inhalation
Glacial acetic acid—corrosive to skin and tissue, toxic by ingestion
Sulfuric acid—corrosive to skin eyes and tissue
Sodium hydroxide—corrosive, may cause skin burns
Phenolphthalein—no health risks

Procedure

CAUTION! CONCENTRATED SULFURIC ACID AND GLACIAL ACETIC ACID ARE DANGEROUS. IF YOU SPILL EITHER ON YOURSELF, WASH IMMEDIATELY AND INFORM YOUR INSTRUCTOR. WEAR SAFETY GLASSES AT ALL TIMES!!

NOTE: It requires several days for this reaction to reach equilibrium at room temperature (i.e., the rates of both the forward and reverse reactions are slow).

You will therefore START the experiment THIS WEEK, and then will FINISH the experiment in the NEXT lab period. THIS WEEK, you will mix the reagents as described below and complete steps (1), (2), and (3) of the procedure outlined. The SECOND LAB PERIOD, you will complete the procedure in this experiment before going on to whatever other work is assigned for the day.

The 0.200 M NaOH solution will be measured out with a piece of apparatus called a BURET. A buret can be used for measuring volumes of solutions very ACCURATELY and PRECISELY (the volume can be read to 0.02 mL). With such precision possible, it is essential that the buret be absolutely CLEAN.

Take out a 400 mL beaker out of your locker and use it to collect waste.

The buret should be washed out with soap and water, then rinsed with distilled water. Finally, to remove any distilled water remaining in the buret, the buret should be washed out with several 5 to 10 mL portions of the solution which is to be used in the buret. (**The rinsing should be DISCARDED into your waste beaker.**)

Obtain no more than 100 mL of 0.200 M NaOH solution in a clean, dry beaker. Clean, rinse, and fill the buret with the 0.200 M NaOH solution and set aside until needed below.

Note: Use a graduated cylinder for the following measurements. Do not place acetic acid or isopropyl alcohol into a buret. They will cause the plastic to dissolve.

In a 250 mL Erlenmeyer flask combine 14.2 mL of glacial acetic acid (CAUTION) with 19.2 mL of isopropyl alcohol. These volumes correspond to equal moles of both substances. **This mixture will be called the REACTION MIXTURE.**

Mix thoroughly and stopper with a rubber stopper covered with plastic wrap (to prevent fumes from the reaction mixture from attacking the rubber).

Proceed immediately to the following parts.

(1)   Remove 1 mL of the reaction mixture with a pipet (USE A RUBBER BULB OR A PIPET PUMP, **NOT YOUR MOUTH**), and use the mixture to rinse the pipet.

Discard this solution into the waste beaker you took out of your locker.

Measure out 25 mL of distilled water and pour it into a 125 mL Erlenmeyer flask. Pipet another 1 mL of the reaction mixture into the 25 mL of distilled water in your Erlenmeyer flask.

Add three drops of phenolphthalein indicator, and titrate with 0.200 M NaOH from the buret to a pale pink endpoint. Do not overshoot this end point—you can't go back.

Discard this solution into the waste beaker at your work station.

Record the volume of NaOH used in your lab notebook. The results of this titration will enable you to calculate the initial concentration of acetic acid in the reaction mixture, and from this, the concentration of isopropyl alcohol also.

To the **reaction mixture** (the flask that contains the acetic acid/isopropyl alcohol mixture) add exactly 15 drops of CONCENTRATED SULFURIC ACID catalyst. (CAUTION! BURNS SKIN—AVOID HANDLING THE BOTTLE WITH YOUR FINGERS—WASH HANDS AFTER USING!!).

Stopper the flask. Then put it in your locker for use next week.

(2)   Measure out 31.5 mL (measured with a graduated cylinder) of distilled water and pour it into a clean Erlenmeyer flask.

Add exactly 15 drops of concentrated sulfuric acid (CAUTION) to this (if by chance, you used more or less than 15 drops of sulfuric acid to catalyze the reaction mixture above, use that amount here also).

After stirring the sulfuric acid/water solution, use several 1 mL portions of this solution to rinse the 1 mL pipet.

Discard these rinses into your waste beaker.

Rinse out an empty 125 mL Erlenmeyer flask and pour 25 mL of distilled water into it (use your graduated cylinder to measure).

Next, pipet exactly 1 mL of the sulfuric acid/water solution into this 25 mL of distilled water.

Add 3 drops of phenolphthalein solution, and titrate with 0.200 M NaOH to a pale pink endpoint. Record the volume of NaOH used. The volume of NaOH used in this step is the correction to be applied for sulfuric acid to the volume of NaOH to be used in step (4) below on the equilibrium mixture.

Discard the contents of this flask into your waste beaker.

At this point check your results with your lab instructor. If your results are satisfactory, drain any leftover NaOH in your buret into the waste beaker, if not repeat step (1) or step (2) as directed.

(3)   If you have not already done so, place the stoppered Erlenmeyer flask containing the acetic acid/isopropyl alcohol mixture (the reaction mixture) into your drawer until the next lab period.

Discard the contents of your waste beaker into the WASTE EQUILIBRIUM CONSTANT ACID/BASE bottle.

You will perform step (4) next week.

(4)   **Again take a 400 mL beaker out of your locker and use it to collect your waste.**

After the reaction mixture has stood for several days, pipet out a small amount (to rinse the pipet).

Discard these rinses into your waste beaker.

Then pipet exactly 1 mL of the mixture into about 25 mL of distilled water.

Obtain approximately 75 mL of 0.200 M NaOH solution, and rinse and fill the buret with this solution.

Add 3 drops of phenolphthalein indicator to the 1 mL sample of equilibrium reaction mixture you have diluted with water, and titrate with standard 0.200 M NaOH to the pale pink end point. Record the volume of NaOH used in this step.

This amount of NaOH used (for the combination of acetic acid and sulfuric acid in the equilibrium mixture) must be corrected for the amount of sulfuric acid present—from step (2) above—before the concentration of acetic acid in the equilibrium mixture can be calculated.

Discard the contents of your 125 mL flask into the waste beaker.

Check your results with your lab instructor. If the results are satisfactory, you have finished the experiment. If they are not, redo the step 4 titration.

Discard the contents of your 125 mL Erlenmeyer flask and any leftover NaOH into your waste beaker.

Discard the contents of your waste beaker into the WASTE EQUILIBRIUM CONSTANT ACID/BASE bottle.

Discard the contents of your 250 mL Erlenmeyer flask (the one that was stoppered and left in your locker) into the WASTE EQUILIBRIUM CONSTANT ACETIC ACID/ISOPROPYL ALCOHOL/ISOPROPYL ACETATE bottle.

REMEMBER TO TURN IN THE YELLOW COPY OF YOUR LAB NOTEBOOK TO YOUR TA.

Le Chatelier's Principle and Equilibrium

Introduction

Many reactions come to EQUILIBRIUM. In the reaction you have been studying during the last two lab periods, the reaction seemed to have STOPPED before the full amount of product expected had been formed. When equilibrium had been reached in this system, there were significant amounts of both products as well as original reactants still present. In today's experiment, you will study CHANGES made in a system ALREADY IN EQUILIBRIUM.

Le Chatelier's Principle states that, if we DISTURB a system that is ALREADY IN EQUILIBRIUM, then the system will REACT so as to MINIMIZE THE DISTURBANCE. This is most easily demonstrated in cases where ADDITIONAL REAGENT IS ADDED to a system in equilibrium, or when ONE OF THE REAGENTS IS REMOVED from the system in equilibrium.

SOLUBILITY EQUILIBRIA

Suppose we have a solution that has been SATURATED with a solute: This means that the solution has already dissolved as much solute as possible. If we try to dissolve ADDITIONAL SOLUTE, no more will dissolve, because the saturated solution will be IN EQUILIBRIUM with the SOLUTE.

$$SOLUTE + SOLVENT \rightleftharpoons SOLUTION$$

Le Chatelier's Principle is most easily seen when an IONIC solute is used. Suppose we have a saturated solution of sodium chloride, NaCl. Then the reaction

$$NaCl(s) \rightleftharpoons Na^+(aq) + Cl^-(aq)$$

will describe the equilibrium that exists. Suppose we then try adding an additional amount of one of the ions in solution. For example, suppose we added several drops of HCl solution (which contains the chloride ion). According to Le Chatelier's Principle, the equilibrium would shift so as to CONSUME some of the added chloride ion. This would result in a net DECREASE in the amount of NaCl that could dissolve. If we watched the saturated NaCl solution as the HCl was added, we should see some of the NaCl PRECIPITATE because less could dissolve when an additional chloride ion was added.

COMPLEX ION EQUILIBRIA

Oftentimes, dissolved metal ions will react with certain substances to produce brightly colored species called COMPLEX IONS. For example, iron(III) reacts with the thiocyanate ion (SCN^-) to produce a bright RED complex ion:

$$Fe^{3+} + SCN^- \rightleftharpoons [FeNCS^{2+}]$$

This is an equilibrium process that is easy to study, because we can monitor the bright red color of $[FeNCS^{2+}]$ as an indication of the position of the equilibrium. If the solution is very red, there is a lot of $[FeNCS^{2+}]$ present; if the solution is NOT very red, then there must NOT be a lot of $[FeNCS^{2+}]$ present.

Using this equilibrium, we can try adding additional Fe^{3+} or additional SCN^- to see what effect this has on the red color according to Le Chatelier's Principle. We will also add a reagent that REMOVES SCN^- from the system to see what effect this has on the red color.

ACID-BASE EQUILIBRIA

Although you have not yet studied the properties of acids and bases in detail, you will learn that many acids and bases exist in solution in equilibrium sorts of conditions. For example, the substance ammonia is involved in an equilibrium in aqueous solution.

$$NH_3 + H_2O \rightleftharpoons NH_4^+ + OH^-$$

Once again, we can use Le Chatelier's Principle to study this equilibrium. We will try adding additional ammonium ion or some additional hydrogen ion to see what happens. Since none of the components of this system is itself colored, we will be adding an acid-base indicator, which changes color with pH, to have an index of the position of the ammonia equilibrium. The indicator we will use is the same one that you used last week, phenolphthalein, which is PINK in basic solution and COLORLESS in acidic solution.

CHEMICALS

Iron(III) chloride ($FeCl_3$)—skin and tissue irritant, contains hydrochloric acid for stability so it is corrosive

Potassium thiocyanate (KSCN)—toxic by ingestion

Silver nitrate ($AgNO_3$)—toxic, avoid contact with skin and eyes, stains skin

Ammonia 6.0 M (NH_3)—respiratory irritant

Ammonium chloride (NH_4Cl)—toxic by ingestion

Phenolphthalein—no major health risks

Hydrochloric acid 6.0 M (HCl)—toxic by ingestion, corrosive to skin and eyes

Hydrochloric acid 12.0 M (HCl)—highly toxic by ingestion or inhalation, severely corrosive to skin and eyes

Saturated sodium chloride (NaCl)—no major health risks

Procedure

CAUTION! CONCENTRATED AMMONIA IS A RESPIRATORY IRRITANT—USE IN THE HOOD. WEAR SAFETY GLASSES AT ALL TIMES!!

SOLUBILITY EQUILIBRIA

Obtain 5 mL of saturated sodium chloride solution in a test tube. This solution was prepared by adding solid NaCl to water until no more would dissolve. Undissolved NaCl is left in the bottle to make sure that the solution is saturated.

Add 5 drops of 12 M HCl to the NaCl solution. A small amount of solid NaCl should form and precipitate out of the solution. The crystals may form slowly, and may be very small. Examine the test tube carefully. If a precipitate does not form, add a few more drops of 12 M HCl.

Discard this solution in the WASTE NaCl/HCl bottle.

In your lab notebook, describe what happens in terms of Le Chatelier's Principle

COMPLEX ION EQUILIBRIA

Prepare a stock sample of the bright RED complex ion $[FeNCS^{2+}]$ by mixing 3 drops of 0.1 M iron(III) chloride and 3 drops of 0.1 M KSCN solutions. The color of this mixture is too dark to see color changes as it is, so dilute this mixture with 25 mL of water.

Pour about 5 mL of the diluted red stock solution into each of four test tubes. Label the test tubes as #1, #2, #3, and #4.

Test tube #1 will have no change made in it, so that you can use it for comparisons of color with what will be happening in the other test tubes.

To test tube #2, add about 5 drops of 0.1 M $FeCl_3$ solution.

To test tube #3, add about 5 drops of 0.1 M KSCN solution.

To test tube #4, add $AgNO_3$ solution drop-wise until a change becomes evident. The Ag^+ ion REMOVES the SCN^- ion from the solution as a solid (silver thiocyanate).

Discard all solutions in this part of the experiment into the WASTE $FeCl_3$/KSCN/$AgNO_3$ bottle.

Describe the intensification or fading of the red color in each test tube in your lab notebook in terms of Le Chatelier's Principle.

ACID-BASE EQUILIBRIA

Prepare a dilute ammonia solution by adding 4 drops of concentrated ammonia (6 M) (CAUTION—RESPIRATORY IRRITANT) to 25 mL of water.

Add 3 drops of phenolphthalein to the dilute ammonia solution, which will turn PINK (ammonia is a base, and phenolphthalein is pink in basic solution).

Place about 5 mL of the pink dilute ammonia solution into each of two test tubes.

To one of the test tubes, add several small crystals of ammonium chloride (which contains the ammonium ion, NH_4^+).

To the other test tube, add a few drops of 6 M HCl.

Discard all of the solutions from this part of the experiment into the WASTE NH3/HCl/NH4Cl bottle.

In your lab notebook, describe what happens to the pink color in terms of how Le Chatelier's Principle is affecting the ammonia equilibrium.

REMEMBER TO TURN IN THE YELLOW COPY OF YOUR LAB NOTEBOOK TO YOUR TA.

$$NH_3 + H_2O \longrightarrow NH_4^+ + OH^-$$

$$.025 \cdot 1.19$$
$$.025 \cdot 1.975$$

$$\frac{25 \, mL}{1.975 \frac{Mol}{L}} \qquad \frac{500 \, mL}{x}$$

182

Volumetric Analysis—Acid-Base Titrations

Introduction

VOLUMETRIC ANALYSIS involves the application of the MOLE CONCEPT to reactions of solutes in SOLUTION. In most lab practice, CONCENTRATION of a solution is most conveniently expressed in terms of the solution's MOLARITY, M: the number of moles of solute dissolved per liter of solution.

From the MOLARITY, the amount of solute involved in a reaction in solution can be determined by just measuring the VOLUME of the solution in question. For example, 50.0 mL of 0.100 M salt solution contains (mole = M x L):

$$(0.0500 \text{ liters})(0.100 \text{ moles/liter}) = 0.00500 \text{ moles of salt}$$

Volumetric analysis employs the technique of TITRATION in which the VOLUMES of solutions required to react EXACTLY with each other are determined. The VOLUMES of the reacting solutions are measured carefully with a device called a BURET. The scale divisions on the buret enable you to measure volumes to the nearest 0.01 mL. The point where one solution has been added in just the right amount to react exactly with the other solution present is called the EQUIVALENCE POINT of the titration. The equivalence point is determined visually by use of a substance called an INDICATOR (in acid-base titrations, such as in this experiment, the indicator used is a dye that CHANGES COLOR if the solution turns from slightly acidic to slightly basic).

In this experiment, you will be given an UNKNOWN SOLUTION containing a certain concentration of ACETIC ACID, CH_3COOH. We will determine the CONCENTRATION of acetic acid present, by titrating the UNKNOWN acetic acid solution with a solution of SODIUM HYDROXIDE (NaOH) of KNOWN concentration. Acetic acid and sodium hydroxide react according to

$$CH_3COOH + NaOH \rightarrow CH_3COONa + H_2O$$

i.e., in a one-to-one ratio. Suppose we find it takes 25.02 mL of 0.1000 M NaOH solution to titrate 22.09 mL of an unknown acetic acid solution. The number of moles of NaOH that have been used in the titration is

$$\text{moles NaOH} = 0.02502 \text{ liters} \times 0.1000 \text{ moles/liter} = 0.002502 \text{ moles NaOH}$$

But, since the chemical reaction has 1:1 stoichiometry, there must have been 0.002502 moles of acetic acid in the unknown solution sample. Therefore, the CONCENTRATION in moles/liter, of the acetic acid in the unknown must be

$$M_{\text{Acetic Acid}} = \frac{0.002502 \text{ mole}}{0.002209 \text{ Liters}} = 0.1133 \text{ M}$$

There is one minor complication to this experiment that we have to get around, however. SODIUM HYDROXIDE SOLUTIONS ARE NOT VERY STABLE (they react with the glass of the bottles used to store them, and also can absorb carbon dioxide gas from the air). So we can't provide you with a solution of NaOH whose concentration is known precisely enough for the accuracy required in this sort of experiment. You will have to prepare your own solution of NaOH, and then DETERMINE its concentration accurately and precisely.

Sodium hydroxide also reacts on a 1:1 stoichiometric basis with HYDROCHLORIC ACID, HCl

$$NaOH + HCl \rightarrow NaCl + H_2O \quad \text{(net ionic: } H^+ + OH^- \rightarrow H_2O\text{)}$$

Luckily, solutions of hydrochloric acid can be made up conveniently to KNOWN CONCENTRATIONS, and can be KEPT STORED for long periods of time without significant decomposition. In this experiment, you will be provided with a STANDARD solution of exactly 0.1000 M HCl solution. You will prepare a solution of sodium hydroxide of APPROXIMATELY 0.1 M, and then, by titrating the NaOH solution with the STANDARD HCl solution, you will determine the EXACT concentration of the NaOH solution (to four significant figures). The calculations involved are completely analogous to the calculation described above for the unknown acetic acid. Once you know the concentration of your own NaOH solution accurately, you can then use this solution to titrate the acetic acid unknown in order to determine its concentration.

In addition to determining the concentration of the acetic acid unknown solution in terms of its MOLARITY, we also want to report the amount of acetic acid present in the unknown solution as the PERCENT acetic acid in the solution BY WEIGHT. To convert from the MOLAR CONCENTRATION (of a solution) to a PERCENTAGE (of solute by weight), the DENSITY of the solution is needed for your unknown (which is mostly water). You can take the density of the solution to be 1.00 g/mL. Additionally the FORMULA WEIGHT of the solute (CH_3COOH) is equal to 60.0 g/mole.

Suppose we do a titration as described above, and determine that our acetic acid unknown has a concentration of 0.1133 moles/liter. The following would be how we could convert this MOLARITY to PERCENT BY WEIGHT. Notice how the units cancel:

$$\frac{0.1133 \text{ mole Acetic Acid}}{1 \text{ Liter solution}} \times \frac{1.000 \text{ Lsolution}}{1000 \text{ mL solution}} \times \frac{1.000 \text{ mL solution}}{1.00 \text{ g solution}} \times \frac{60.0 \text{ g acetic acid}}{1 \text{ mole acetic acid}} \times 100 = 0.68\%$$

(Notice that the units are "g acetic acid/g solution.") The factor of 100 converts the FRACTION to PERCENT (per 100). Numerically, the conversion from molarity to percent by weight is trivial, since many of the numbers in the above conversion cancel out.

Summary

Approximately 0.1 M NaOH solution is prepared by dilution of concentrated stock solution (2.0 M). The approximately 0.1 M NaOH is standardized by titration with standard 0.1000 M HCl solution using phenolphthalein as an indicator (this indicator is colorless in acidic solution, but pink in basic solution). The now known NaOH solution is then used to determine the concentration of an unknown acetic acid solution by a similar titration. The percent by weight of acetic acid in the unknown is calculated.

Supplies

Two burets and clamp; stock NaOH solution; standard 0.1000 M HCl solution; phenolphthalein indicator solution; acetic acid unknown. RECORD THE NUMBER OF THE UNKNOWN IMMEDIATELY IN YOUR LAB NOTEBOOK.

CHEMICALS

Hydrochloric acid (HCl)—toxic and corrosive, will burn skin
Sodium hydroxide (NaOH)—toxic and corrosive, easily absorbed through the skin
Acetic acid—corrosive to skin and tissue, can cause skin burns
Phenolphthalein—no major health risks

Procedure

NO SOLUTIONS FROM THIS EXPERIMENT SHOULD GO DOWN THE DRAIN.

CAUTION! WEAR SAFETY GLASSES AT ALL TIMES!!

(1) Preparing the burets

For accurate results in any titration, the burets used must be scrupulously CLEAN. Water (or titrating solutions) should flow IN SHEETS down the inside of the buret, and should NOT ADHERE in drops to the inside of the glass.

Obtain two burets and a buret clamp from your labroom. Rinse the burets thoroughly with water to remove any previous chemicals. If water adheres to the interior of the buret, obtain some soap solution and a special "buret brush" from the stockroom, and scrub the buret until water no longer beads up. Rinse finally with water.

Before a buret is subsequently filled with any solution, THE BURET SHOULD BE RINSED WITH THREE OR FOUR 5 to10 mL PORTIONS OF THE SOLUTION THAT ULTIMATELY IS TO BE PLACED IN THE BURET. This makes sure that the solution to be placed in the buret will not be diluted inadvertently by any residual water left in the buret from the washing.

(2) Preparation of solutions and filling the burets

Obtain 25 mL of 2.0 M NaOH solution, and add this to 475 mL of distilled water in your Florence (or 500 mL Erlenmeyer) flask.

Stir the solution thoroughly. This will produce 500 mL of approximately 0.10 M NaOH solution.

Stopper the flask when not using the solution (to keep the NaOH from reaction with carbon dioxide in the air).

Thoroughly CLEAN and DRY a 250 mL beaker, and obtain around 75 mL of 0.1000 M HCl standard solution. Keep the beaker covered with a watch glass during the experiment (the beaker must be clean and dry, and must be kept covered, to prevent the concentration of the HCl solution from changing).

Rinse one buret several times with small portions of your dilute NaOH solution. **Discard the rinsing into a 400 mL beaker from your locker to be used for waste**. Finally fill the buret with the dilute NaOH solution.

Rinse the other buret several times with small portions of the standard HCl solution **Discard the rinsing into a 400 mL beaker from your locker to be used for waste**. Finally fill your buret.

Make certain there are no air bubbles trapped in the tip of either buret. (Ask your lab instructor if you can't get the bubbles out yourself.)

Record (in your lab notebook and on the report sheet) the initial readings of both burets, reading across the bottom of the curved liquid surface (called the MENISCUS). Make your readings to TWO DECIMAL PLACES (any less precision will give meaningless results).

NOTE: Initial readings should be reported as zero mL (or close to it), not 50 mL.

(3) Standardization of NaOH solution

Run around 20 to 25 mL of the HCl standard solution from the buret into a clean 250 mL Erlenmeyer flask, and add 2 or 3 drops of the phenolphthalein indicator solution. It is NOT necessary to read the burets at this point. The 20 to 25 mL of solution run out at the start will be INCLUDED in the reading taken at the end point (color change) of the titration.

Titrate the HCl solution by adding NaOH solution from the other buret until the solution in the Erlenmeyer flask just turns pink (keep swirling the flask while adding the NaOH). If the pink color does not remain for at least 30 seconds, add another drop of NaOH until it does stay pink for at least 30 seconds.

The appearance of the pink color means you have added MORE NaOH than is required to react with the HCl in the flask.

Carefully "back titrate" the solution in the flask by adding HCl from the other buret drop by drop (swirling the flask constantly), until ONE SINGLE DROP of HCl from the buret causes the pink color to JUST disappear.

Confirm that you are now at the end point by adding ONE SINGLE drop of NaOH from the buret (careful!); the pink color should just return faintly.

Record the readings of the burets (again reading the bottom of the meniscus) to the nearest 0.01 mL. (Final readings should be read somewhere between 20 and 25 mL.)

Discard this solution into your waste beaker. When the waste beaker becomes filled, discard the contents into the WASTE ACID-BASE container.

From the volume of each solution used, and from the known molarity of the HCl, calculate the molarity of the NaOH solution.

Refill the burets and repeat the entire process above.

Calculate the molarity of the NaOH from the results of this second titration. If the two molarities of NaOH do not agree within 1%, repeat the standardization a third time or until two molarities are obtained within 1% of each other.

Use the AVERAGE of your two closest molarities as the concentration of your NaOH solution.

Note: It is not necessary to always fill a buret to the zero mark (especially if your height prevents you from seeing the zero mark easily). If the actual initial volume reading is recorded, as well as the final volume reading (at the color change), then the volume of solution used is just (Vol$_{final}$ - Vol$_{initial}$).

(4) Analysis of the acetic acid unknown

Drain the buret that contained HCl into your waste beaker, rinsing it several times with distilled water. Discard the rinses also into your waste beaker.

Then rinse the buret with several small portions of your acetic acid unknown. **Again, discard the rinses into your waste beaker.** Finally fill the buret with the unknown, again making sure to avoid air bubbles.

Run about 20 to 25 mL of the unknown from the buret into a clean 250 mL Erlenmeyer flask, add 2 or 3 drops of phenolphthalein solution, and titrate with the standard NaOH to the end point.

Discard this solution into your waste beaker.

Based on the volumes of NaOH and unknown used, and on the now known molarity of the NaOH, calculate the molarity of the unknown acetic acid solution.

Repeat the titration of the unknown until you get two calculated molarities of the unknown within 1% of each other. Use the average of your two best values as the concentration of your unknown.

Discard the titration solutions, any leftover solutions in your burets into your waste beaker. Transfer the contents of your waste beaker and any leftover NaOH (in your Florence flask or 500 mL Erlenmeyer flask) into the WASTE ACID-BASE container.

Finally, calculate the percent by weight of acetic acid in your unknown. Since your unknown solution is quite dilute, you are valid in making the approximation that its density is 1.00 g/mL; the molecular formula of acetic acid is CH_3COOH, which means a formula weight of 60.0g/mole.

REMEMBER TO TURN IN THE YELLOW COPY OF YOUR LAB NOTEBOOK TO YOUR TA.

Solubility Product of Calcium Hydroxide

INTRODUCTION

Solubility is defined as the maximum amount of solute that can be dissolved in a given amount of solvent. Usually we express solubility in terms of the grams of solute that can be dissolved in 100 mL of water at a given temperature. The amount of solute that can be dissolved in a given quantity of solvent is not an unlimited quantity. We also have learned how to use the solubility rules to determine whether a substance is soluble or insoluble in water.
In general, we will consider a solute to be soluble if as little as 0.01 mole of solute can be dissolved in a liter of solvent. Substances in which less than this quantity can be dissolved are considered to be insoluble. We should recognize that for a substance with a molar mass of 100 g/mole, that only one gram of this substance could be dissolved in a liter of solution and the substance would be considered soluble. Any substance that has a solubility less than 0.01 mole per liter is considered to be insoluble.

We also have learned that for any substance the solubility process represents a dynamic equilibrium. That is, the rate of dissolving is equal to the rate of crystallization. In other words, even for insoluble substances, a small amount of the solid will dissolve in water. For example, silver chloride (AgCl) is considered to be insoluble in water. However, at 25°C, 1.34×10^{-5} mole (about 0.00193 g) of AgCl can be dissolved in a liter of water. We can represent this process as follows:

$$AgCl(s) \rightleftharpoons Ag^+(aq) + Cl^-(aq)$$

Since this is an equilibrium process, we can write an equilibrium expression:

$$K_c = [Ag^+][Cl^-]$$

Since this process represents the dissolving of the small amount of AgCl in water, we give this K a special subscript. It is called K_{sp}, where the sp represents the solubility product. Note, as expected since AgCl is a solid, it is not included in the equilibrium expression (the concentration of a solid is constant and this value is included in the equilibrium constant).

$$K_{sp} = [Ag^+][Cl^-]$$

From the stoichiometry we can see that if we started with pure AgCl, the concentration of the ions Ag^+ and Cl^- should be equal and the concentration of either dissolved ion will represent the concentration of AgCl that dissolved. Since we are multiplying the concentration of both ions, the result of this operation represents the product of the solubility squared. Thus, the term solubility product is used. For silver chloride, the K_{sp} value is 1.8×10^{-10}. That is,

$$K_{sp} = [Ag^+][Cl^-] = 1.8 \times 10^{-10}$$

As for any equilibrium process, we can set up an ICE Table for this reaction as follows:

We start with solid AgCl and at the instant we add the solid to water, there are no ions present. At equilibrium, the amount of both the silver ions and the chloride ions must be equal, so we represent the concentration of each as x.

Plugging in the data, we get:

$$K_{sp} = [Ag^+][Cl^-] = 1.8 \times 10^{-10}$$
$$K_{sp} = [x][x] = 1.8 \times 10^{-10}$$
$$1.8 \times 10^{-10} = [x]^2$$

so
$$x = \sqrt{1.8 \times 10^{-10}} = 1.34 \times 10^{-5} \text{ M}$$

The solubility of AgCl in water at 25°C is 1.34×10^{-5} M.

For a buffer solution, we determined that the addition of an ion common to the equilibrium caused the hydrogen ion concentration to decrease. The same is true for slightly soluble salts. The presence of a common ion results in a decrease in the solubility of the salt.

For example, let's consider a solution of some NaCl (a soluble salt) to which we will add some solid AgCl. If the solution originally contained 0.1 M NaCl, then the concentration of both the sodium ions and chloride ions must be 0.1M (the solubility of NaCl is considerably greater than 0.1 mole per liter). The equilibrium process for AgCl has not changed except the solution now initially contains 0.1 M Na^+ ions and 0.1 M Cl^- ions. Since sodium ions are not involved in this equilibrium, we do not need to concern ourselves with them. Let's do a new ICE Table for this system:

$$AgCl(s) \rightleftharpoons Ag^+(aq) + Cl^-(aq)$$

I	S	0	0.1
	O		
C	L	+x	+x
	I		
E	D	x	0.1 + x

We can assume that the x is small compared to 0.1 M and can be ignored. That is,

$$0.1 + x = 0.1$$

$$K_{sp} = [Ag^+][Cl^-] = 1.8 \times 10^{-10}$$
$$K_{sp} = [x][0.1] = 1.8 \times 10^{-10}$$
$$1.8 \times 10^{-10} = [x][0.1]$$
$$[x] = 1.8 \times 10^{-10}/[0.1] = 1.8 \times 10^{-9} \text{ M}$$

where x again represents the solubility of AgCl.

We can see that the solubility of AgCl has dropped from 1.34×10^{-5} M to 1.8×10^{-9} M (about 2.58×10^{-7} g/liter).

We can also explain this effect by using Le Chatelier's Principle. In this case, let's add the NaCl to an equilibrium mixture that contains dissolved AgCl. If we add chloride ions to the equilibrium, the reaction must shift to use up some of the added chloride ions. This means the reaction shifts to make more solid AgCl, thereby reducing the solubility of the AgCl. (The instant we add the chloride ions $Q > K_{sp}$, so the reaction shifts to make more solid AgCl.)

CALCULATIONS

In this experiment we will determine the molar solubility and the K_{sp} value for calcium hydroxide. Then we will determine the solubility of a solution of calcium hydroxide to which some additional calcium ions have been added (common ion effect).

Determination of the molar solubility of $Ca(OH)_2$ and its K_{sp} value

The equilibrium is:

$$Ca(OH)_2(s) \rightleftharpoons Ca^{2+}(aq) + 2OH^-(aq)$$

We will start with a saturated solution of calcium hydroxide (one in which the maximum amount of calcium hydroxide has been dissolved) and titrate the solution with hydrochloric acid of known molarity. The balanced equation for this titration is:

$$Ca(OH)_2 + 2HCl \rightarrow CaCl_2 + 2H_2O$$

Using the known molarity of the HCl and the volume of HCl, we can find the mole of HCl (mole HCl = M x V_{HCl}). Then using the coefficients of the balanced equation we can determine the concentration of the dissolved calcium hydroxide in the solution. That is, this concentration corresponds to the molar solubility of calcium hydroxide. Next, we need to determine the K_{sp} value for calcium hydroxide. The process for dissolving calcium hydroxide is:

$$Ca(OH)_2(s) \rightleftharpoons Ca^{2+}(aq) + 2OH^-(aq)$$

The K_{sp} expression is

$$K_{sp} = [Ca^{2+}][OH^-]^2$$

From the above equation we see that the concentration of the calcium ions is the same as the molar solubility of $Ca(OH)_2$ because they are in a 1:1 molar ratio. We can also determine the concentration of the hydroxide ions using the coefficients of the balanced equation.

$$[OH^-] = 2 \times [Ca^{2+}]$$

Once we know the concentration of both the calcium ion and the hydroxide ion, we can simply plug them back into the K_{sp} expression and then calculate K_{sp}.

Determination of the solubility of calcium hydroxide in a solution that contains some calcium ions:

The solution used in this part contains a saturated solution of calcium hydroxide to which 11.1 g of calcium chloride ($CaCl_2$) has been added per liter of solution.

However, the calculations are similar to part 1.

The net ionic equation for the reaction that occurs between hydrogen ions and hydroxide ions is simply:

$$H^+ + OH^- \rightarrow H_2O$$

First, find the mole of HCl as before. Next determine the concentration of the hydroxide ions using the above net ionic equation. (From the balanced equation, we see that the concentration of HCl equals the concentration of hydrogen ions and the concentration of hydrogen ions equals the concentration of hydroxide ions.)

Lastly, use the equation $Ca(OH)_2(s) \Leftrightarrow Ca^{2+}(aq) + 2OH^-(aq)$ to calculate the concentration of calcium hydroxide that has dissolved. This is the molar solubility of calcium hydroxide.

Molar solubility of calcium hydroxide = ½ $[OH^-]$

CHEMICAL RESPONSIBILITY

Calcium hydroxide (saturated)—corrosive, skin irritant
Hydrochloric acid—corrosive
Methyl orange indicator—toxic by ingestion
Calcium chloride—no major health risks

PROCEDURE

Determination of the molar solubility and K_{sp} of $Ca(OH)_2$

Note: Use a 400 or 600 mL beaker for collecting your chemical waste.

Obtain a 50 mL buret from the hood in your room. Clean it with tap water and then rinse it with two 10 mL portions of distilled water. Next into a 250 mL beaker pour about 100 mL of 0.100 M HCl solution from the hood in your lab room. Rinse the buret with two 5 to 10 mL portions of this solution. Drain this into your waste beaker. Fill your buret with the HCl solution. Make sure to record the initial volume (about 0 mL to two decimal places as best you can) on your report sheet.

Obtain about 100 mL of the saturated calcium hydroxide solution and a 25 mL graduated pipet from your lab room. Rinse the pipet twice with a 2 to 3 mL portion of the saturated calcium hydroxide solution. Discard the solution into your waste beaker. Pipet 25 mL of the solution into a 125 mL flask and add 2 drops of the methyl orange indicator.

Titrate the solution with the HCl solution until the solution changes color from yellow to a light red (or an orange-red). Make sure to swirl the flask continuously during this addition and red color should remain for at least 30 seconds. You cannot overshoot this endpoint, so make sure to do this first titration slowly. Pour this solution into your waste beaker. Rinse the flask with tap water and once with a small amount of distilled water. These rinses can go down the drain.

If needed, refill your buret with HCl. Pipet another 25 mL sample of the saturated calcium hydroxide into your 125 mL flask, add 2 drops of indicator and titrate to the light red end point. This time you should know about how much HCl the neutralization should require so you can quickly allow HCl to be added until your volume of HCl added is within about 1 mL of the first titration. Pour this solution into your waste beaker. Rinse the flask with tap water and once with a small amount of distilled water. These rinses can go down the drain.

Repeat this procedure once more. Check with your lab instructor to make sure your results are acceptable.

Discard any remaining calcium hydroxide into your waste beaker.

Pour the contents of your waste beaker into the 5 gallon acid-base waste container in the hood.

The Molar Solubility of Ca(OH)$_2$ in the Presence of Ca^{2+} Ions

Refill your buret with hydrochloric acid solution.

Obtain about 100 mL of the solution labeled saturated calcium hydroxide/calcium chloride.

Repeat the above procedure for this solution. Again perform three titrations and check with your instructors to make sure your results are okay.

Discard any remaining calcium hydroxide into your waste beaker.

Pour the contents of your waste beaker into the 5 gallon acid-base waste container in the hood.

$$10.\cancel{8}5 \ mL \cdot \frac{1 \ L}{1000 \ mL} = 0.01045 \ L \ HCl \times \frac{0.1 \ mols}{1 \ L} = 0.00\cancel{0}\text{o}\ mols\ HCl$$

$$0.00\cancel{0}1045 \ Hl = OH^- \cdot \frac{1 \ Ca(OH)_2}{2 \ OH}$$

Redox Titration: Analysis of Bleaches

Like acid-base reactions, oxidation-reduction reactions can be used as an analytical tool. In this experiment, we will determine the percentage of hypochlorite ion in laundry bleaching solutions. In liquid bleaching solutions, this ion is present in the form of sodium hypochlorite. In solid bleaches, the ion is more commonly present in the form of calcium hypochlorite. There are also non-chlorine containing bleaches available at the supermarket. The active ingredient in these products is hydrogen peroxide.

The effectiveness of a bleach solution is related to its ability to oxidize substances. Detergents help to remove dirt and grease from clothes by both an emulsification process and agitation. Most liquid bleaches use the hypochlorite ion (usually present in the form of sodium hypochlorite originally) as an oxidizing agent.

Commercial bleaching solutions are made by reacting chlorine gas with the hydroxide ion. If the base is sodium hydroxide, the resulting substance is sodium hypochlorite. Bleaching solutions remove colors from a fabric because of an interaction between the hypochlorite ion and a colored species.

The general process (that occurs in acidic solutions) is as follows:

The hypochlorite ion is converted into the chloride ion

$$ClO^-(aq) \rightarrow Cl^-(aq) \qquad\qquad\qquad (eq\ 1)$$

The iodide ion is converted to elemental iodine (I_2)

$$I^-(aq) \rightarrow I_2(aq) \qquad\qquad\qquad (eq\ 2)$$

The balanced net ionic equation is

$$ClO^-(aq) + 2I^-(aq) + 2H^+ \rightarrow I_2(aq) + Cl^-(aq) + H_2O(l) \qquad\qquad (eq\ 3)$$

This reaction goes to completion, meaning that all of the hypochlorite ions have completely reacted with the iodide ion.

Typically we add some KI solution to distilled water and then add a strong acid. At this point the solution is likely to have a yellowish color. Next, a known volume of bleaching solution is added to the reaction vessel. This results in the immediate formation of elemental iodine (I_2) and the solution turns a deep dark red color.

You may ask how we can measure the amount of hypochlorite ions from this process. The answer is that we cannot determine the amount of hypochlorite from this process. We do know that we added excess iodide ion to a known quantity of the hypochlorite ion. We assume that the iodine formed can be used as a measure of the amount of hypochlorite. There is a very simple and easy reaction that we can perform to determine the amount of elemental iodine formed. It is in fact a reaction that we have previously used in the kinetics experiment. It is to react elemental iodine with the thiosulfate ion as follows:

$$I_2(aq) \;+\; 2\,S_2O_3^{2-}(aq) \;\rightarrow\; 2I^-(aq) \;+\; S_4O_6^{2-}(aq) \qquad\qquad \text{(eq 4)}$$

red-brown   colorless   colorless   colorless

In this reaction the elemental iodine is converted back into the iodide ion.

The next step will be to titrate the iodide solution with a known concentration of sodium thiosulfate. As we add the thiosulfite ion, the amount of I_2 constantly decreases. This is observed by noticing that the very dark red-brown color becomes reddish, then becomes orange, then yellow, and finally colorless. It is very difficult to see when the yellow color has vanished so determining the end point becomes somewhat a guess. However, if we add some starch to the light yellow solution, we can make a much more stunning end point. In the kinetics experiment you performed, the solution eventually turned blue-black because elemental iodine was produced. In this reaction we can perform the reverse process where iodine is being used up and colorless iodide ion is being formed. When we add a small amount of starch to the light yellow iodide solution, it turns dark blue-purple. As the thiosulfate solution is added, the color becomes lighter eventually become a light bluish purple color. At this point it will require only one or two more additional drops of thiosulfate solution to turn the reaction mixture colorless.

NOTE: If the starch solution is added too early in the titration, the formation of the blue-black substance is not easily reacted causing the end point to occur very slowly and making it difficult to observe.

We can now determine the amount of hypochlorite ion in the commercial bleaching solution. From the known concentration of the sodium thiosulfate solution and the volume of this solution we can determine the moles of sodium thiosulfite used in the titration. Using equation 4 we can calculate the mole of elemental iodine formed. Then using equation 2, we can find the moles of hypochlorite ion in the initially solution.

In this reaction it is difficult to know exactly the sodium thiosulfate concentration because a small quantity of the thiosulfate ion undergoes decomposition on mixing with water. A more concentrated version of the thiosulfate solution you will use will be provided to you. You will be required to determine the exact concentration of a diluted thiosulfate solution using a standard solution (a solution whose concentration is known). In this case the standard solution will be potassium iodate. The reaction between the iodate and the thiosulfate ion also will create iodine and will visually appear identical to the process previously described.

The net ionic reaction for this overall process is:

$$IO_3^-(aq) + 6S_2O_3^{2-}(aq) + 6H^+(aq) \rightarrow I^-(aq) + 3S_4O_6^{2-}(aq) + 3H_2O(l)$$

From this reaction, we can see that 6 moles of thiosulfate ions are required for each mole of iodate ions used. Using this information we can convert the concentration of the iodate ions to the concentration of thiosulfate ions.

In this experiment you will first standardize an unknown sodium thiosulfate solution and then use this information to determine the concentration of hypochlorite ion in a commercial bleaching sample.

One possible source of error can arise due to the ability of iodide ions being able to be oxidized to I_2 by oxygen present in the air. The reaction is

$$4 I^- + O_2 + 4H^+ \rightarrow 2 I_2 + 2H_2O$$

This reaction is quite slow in neutral solution. However, the reaction rate increases with acid concentration. To prevent excess I_2 formation that results in extra thiosulfate solution being needed to reach the end point of your titration, one should perform the titration immediately after adding the sulfuric acid solution.

CHEMICAL RESPONSIBILITY

Sodium thiosulfate ($Na_2S_2O_3$)—slightly toxic by ingestion, body tissue irritant
Potassium iodate (KIO_3)—oxidizer, moderately toxic, tissue irritant
Potassium iodide (KI)—no major health risks
Sulfuric acid (H_2SO_4)—corrosive, can cause skin burns
Commercial bleach—corrosive liquid, causes skin burns, moderately toxic by ingestion and
              inhalation
Starch—no major health risks

Caution: Comercial liquid bleaches containing sodium hypochlorite are corrosive to the skin and eyes. If you get any on your skin, wash it off immediately.

PROCEDURE

Step 1:   Standardization of Sodium Thiosulfate Solution

Prepare about 250 mL of an approximate 0.1M solution of sodium thiosulfate by diluting a 1.0 M solution of sodium thiosulfate.

$$2\overset{8}{\cancel{8}}.1$$
$$-1\cancel{3}.9$$
$$\overline{1\ 3\quad 2}$$

$$2\overset{6\ 1}{\cancel{8}}.\cancel{2}0$$
$$-13.75$$
$$\overline{13.45}$$

Obtain a 50 mL buret and clean it out with tap water. Then rinse it with about 15 mL of tap water. These rinses can be discarding into the sink. Use a 400 mL or 600 mL beaker as a waste collection vessel. Lastly rinse it twice with 5 mL portions of your sodium thiosulfate solution. Discard these two rinses in your waste collection beaker. Now you are ready to fill your buret with the sodium thiosulfate solution. To do this, use your gravity funnel. Make sure you do not overfill the funnel or the solution will overflow. You want to add the thiosulfate solution until the liquid level is slightly above the zero mark. Drain down (collecting the waste in your waste beaker) so you are exactly at the zero mark or slightly below it. Read and record (to the nearest 0.02 mL) this volume.

Next obtain a 25 mL pipet and a pipet pump from the supply table in your lab room. Pipet 25 mL of the standard 0.010 M potassium iodate (KIO_3) solution into a clean (it does not need to be dry) 125 mL Erlenmeyer flask. Next, add about 10 mL of 10% KI solution to the flask. Finally add about 5 mL of 1.0 M sulfuric acid to the flask. At this point the solution should have turned a dark red-brown color. Immediately start to titrate this solution with the sodium thiosulfate solution in your buret. As this titration proceeds, the red-brown color will change to red, then to orange, and then to yellow. When the solution has become a very pale yellow, add about 1 mL of 1 % starch (a squirtful) to the solution. This should change the solution color to dark blue. Continue titrating (drop by drop) until the blue color disappears. After each drop is added, swirl your flask. If the blue color persists, add an additional drop and repeat the procedure until the blue color disappears. If you go beyond the end point, you cannot go back in this process.

Discard your waste into your waste beaker.

If needed, refill your buret with additional sodium thiosulfate solution (as described earlier.

Repeat this titration twice more, so that you have three total titrations for the standardization of the sodium thiosulfate solution. Since everybody's dilution will be slightly different, you cannot use your neighbor's data.

Note: All three titrations should yield identical results. So when you do the second and third titrations, you can very quickly add the sodium thiosulfate solution until you get within about 1 to 2 mL of the expected end point. For example, if your first titration required 32.4 mL of sodium thiosulfate, you can very quickly drain in about 31 mL of sodium thiosulfate for the second and third titrations, add your starch, and finish the titration.

After finishing the third titration you can empty your waste beaker into the waste container in the hood (probably a 5 gallon plastic container).

From this data, one can calculate the concentration of your diluted sodium thiosulfate solution.

Step 2: **Titration of Commercial Bleach with Sodium Thiosulfate**

Refill your buret with sodium thiosulfate solution.

Obtain a 1 mL calibrated pipet from the supply table. Next, pour about 10 mL of commercial bleach into your 50 mL beaker (do not use a larger beaker as you will not be able to pipet any liquid out).

Add 25 mL of distilled water to a clean (does not need to be dry) 125 mL or a 250 mL Erlenmeyer flask. Next, add 5 mL of 10% KI solution to the flask and then pipet exactly 1 mL of the commercial bleach solution into the flask. Finally, add about 10 mL of 1.0 M sulfuric acid to the flask. At this point the solution should have turned a dark red-brown color. Immediately start to titrate this solution with the sodium thiosulfate solution in your buret. As this titration proceeds, the red-brown color will change to red, then to orange, and then to yellow. When the solution has become a very pale yellow, add 1 mL (about a squirtful) of 2 % starch to the solution. This should change the solution color to dark blue. Continue titrating (drop by drop) until the blue color disappears. After each drop is added, swirl your flask. If the blue color persists, add an additional drop and repeat the procedure until the blue color disappears. If you go beyond the end point, you cannot go back in this process.

Discard your waste into your waste beaker.

If needed, refill your buret with additional sodium thiosulfate solution (as described earlier.

Repeat this titration twice more, so that you have three titrations for the determination of the percentage of sodium hypochlorite in commercial bleaching solution. Since everybody's concentration of sodium thiosulfate will be slightly different, you cannot use your neighbor's data.

Note: All three titrations should yield identical results. So when you do the second and third titrations, you can very quickly add the sodium thiosulfate solution until you get within about 1 to 2 mL of the expected end point. For example, if your first titration required 32.4 mL of sodium thiosulfate, you can very quickly drain in about 31 mL of sodium thiosulfate for the second and third titrations, add your starch, and finish the titration.

After finishing the third titration, you can empty your waste beaker into waste container in the hood (probably a 5 gallon plastic container).

You can now calculate the concentration and weight percent of sodium hypochlorite in liquid bleaches.

CALCULATIONS

1. **Standardization of sodium thiosulfate**

Using the equation:

$$IO_3^-(aq) + 6S_2O_3^{2-}(aq) + 6H^+(aq) \rightarrow I^-(aq) + 3S_4O_6^{2-}(aq) + 3H_2O(l)$$

You can calculate the concentration of your diluted sodium thiosulfate solution.

From the known concentration of the potassium iodate solution (found on the reagent bottle) and volume used (25 mL), we can calculate the moles of potassium iodate reacted. Using the coefficients of the above balanced equation, we can convert the moles of potassium iodate to moles of sodium thiosulfate. Lastly, using the volume of sodium thiosulfate used in each trial and the moles of sodium thiosulfate calculated above, you can determine the molar concentration of your sodium thiosulfate solutions.

2. **Determining the molar concentration of commercial bleach solution:**

From the calculated concentration of sodium thiosulfate solution calculated above and the volume of sodium thiosulfate used in each titration in part 2, we can determine the moles of sodium thiosulfate needed for each titration.

Then, using the equation:

$$I_2(aq) + 2 S_2O_3^{2-}(aq) \rightarrow 2I^-(aq) + S_4O_6^{2-}(aq) \qquad \text{(eq 4)}$$
$$\text{red-brown} \quad \text{colorless} \qquad \text{colorless} \quad \text{colorless}$$

We can convert the mole of sodium thiosulfate into moles of iodine (I_2).

Note: When sodium thiosulfate dissolves, it produces one mole of thiosulfate ions. That is, the mole of sodium thiosulfate used in each titration is the same as the mole of thiosulfate ion used up.

$$Na_2S_2O_3 \rightarrow 2Na^+ + S_2O_3^{2-}$$

Next, using the equation

$$ClO^-(aq) + 2I^-(aq) + 2H^+ \rightarrow I_2(aq) + Cl^-(aq) + H_2O(l)$$

we can convert the moles of iodine (I_2) into the mole of hypochlorite ions.

Again, when sodium hypochlorite dissolves, we see that the moles of hypochlorite ions must equal the moles of sodium hypochlorite.

$$NaClO \rightarrow Na^+ + ClO^-$$

We have now calculated the concentration of sodium hypochlorite (or the hypochlorite ion in commercial liquid bleaches).

However, most commercial bleaching solutions are reported as a weight percent of sodium hypochlorite. We will need to convert the molarity of sodium hypochlorite into a weight percent.

That is

$$\% \text{ Sodium hypochlorite} = \frac{\text{grams of sodium hypochlorite}}{\text{grams of commercial bleaching solution}} \times 100$$

From the moles of sodium hypochlorite used, we can determine the mass of sodium hypochlorite.

We can find the mass of commercial bleaching solution by using the density of the commercial bleach. This value is 1.084 g/mL. This mass is calculated by

$$\text{mass} = \text{density} \times \text{volume}$$

The volume is the amount of commercial bleach pipetted (1 mL).